# Pha ... ds

A guide to a ... in Egypt

## Supported by:
## MINISTRY OF TOURISM-EGYPT

Dar el Kutub No. 7132/98



Layout & Design by Wafaa Ismail Soliman

Maps by D & B Graphics-Cairo-Egypt

Colour separation by J.C. Center-Cairo-Egypt

Produced by the Printshop of The American University in Cairo Press

First published in 1998 on behalf of the World Bank by

# Contents



**LIST of MAPS &** Main Centres

# Introduction

Welcome to 'Pharaohs' Birds', a guide to help you locate ancient and present-day birds in Egypt. This book covers the entire country which is one of the most popular tourist destinations in the Middle East, with its great history and guaranteed weather. Sadly, at the present time (1997), few people come to find and watch this wildlife, especially the birds which can be seen in their thousands during migration.

This country has a wide variety of sites, with some presently visited by the tourist and some which are not. Most of the tourists are unaware that they share their holiday destination with wildlife. In some cases the wildlife has moved in due to the tourist industry.

Presently, in excess of four million visitors a year come to Egypt and this is estimated to rise to five million in the next few years. It is estimated that only 250 of these actually come to watch the birds! Many of the visitors actually see the birds but often do not know what they are, and the reason for them being there.

Egypt is one of the major crossroads in the world for birds on migration, with an estimated 500 million birds crossing its lands in the autumn and fewer returning during the spring, due to natural losses. Birds can be seen in virtually every square kilometre of Egyptian land but the numbers can vary so much - from one bird to many thousands, depending on the location and on the existing type of habitat.

The rich Nile Valley and Delta areas are good for the people of Egypt as well as for its resident bird-life. Migration can stop over in the country or keep flying/soaring through, using the heat as do many eagles, buzzards, kites, vultures, storks and cranes. Small birds often fly through the night and drop down during the day to feed. Others, like ducks, may winter along the River Nile and on the Delta Lakes while sea birds come to breed and feed in the Red Sea.

This book will take you on a tour of the many sites available to watch birds, while at the same time giving a guide on where to stay and how to get there. It will also delve into the charm of finding the birds of Ancient Egypt with over 75 species to find around the tombs and temples of the Pharaohs.

The book also incorporates a list of the many species of birds already located in Egypt with reference to those seen just over the border in Israel which, due to the large number of birdwatchers visiting there, can number up to 10,000 pairs of eyes in a 2-week period, looking just for birds. With so few birdwatchers on the Egyptian side of the border many birds go unrecorded. The book will also help to highlight the best times of the year for finding birds and, hopefully, act **as a reason for you to visit, or even return to, Egypt!**

## LIST OF PAINTINGS by Mike Henry [mh] & PHOTOGRAPHS by:

John Miles [jm], Keith Temple [kt]
Andrew Heatherington [ah]
Andy Benham [ab]

### Pharaohs' Birds



### Cairo Area



### Port Said

## Aswan



## Abu Simbel



## South Sinai



## Greyscale Paintings

(From colour paintings)

# The Land of Egypt

Most people think of Egypt as a country of desert. This is near to the truth, but it is the magic of diversification that can make even a desert look different. The position on the map can determine whether a country is going to receive a lot of rain or none at all and in this case, will it receive a lot of bird migration? Is the sea water cold or hot? What are the prevailing wind directions? What land adjoins your boundaries? Is there enough food available to keep migrants as winter birds, and are conditions suitable to make them stay to breed?

Many of these questions are interesting and even more so if the country is spending all its time on creating a habitat for just one species of bird as with the heather moorlands for red grouse in Britain. When an area of land is left to its own devices then nature can turn it into an amazing situation given the factors such as water, soil and the right temperature to choose from. Take away or change even one of these and even nature can find it very hard to respond, especially if it is working against the local human population as in overgrazing, the disturbance of the soils and the pollution of the water.

This brings us to the position of the country. Egypt lies in the north eastern part of the African continent. It is bordered by Libya to the west and Sudan to the south. The Red Sea covers most of the east with the Gulf of Aqaba making up the greatest part of the north eastern boundary with Israel, finishing with Gaza right in the very corner. The Mediterranean is the sole boundary in the north. As much as 2,900 kilometres of Egypt's perimeter has a sea boundary.

The country sits next to the Great Rift Valley running northwards from Lake Malawi and the Zambezi River through Tanzania, Kenya, the Ethiopian Highlands and both sides of the Red Sea with the fingers of both the Gulf of Aqaba and the Gulf of Suez. These moving plates have pushed up the mountains of the Eastern Desert and of the Sinai.

The hottest period in the country is from May to September while the coldest periods are during December to February. Most of the rainfall occurs in winter along the Mediterranean with up to 200 millimetres. Cairo can have as little as 28 millimetres a year while Aswan may have 3 millimetres and Siwa 1 millimetre a year! The temperatures can range from 37 centigrade on the coast to 45 centigrade inland in summer but even here the night-time temperature can fall to 0 centigrade in winter. The seas are warm compared with water areas such as the North Sea!

## Geology

This is a brief account of the land and the geology of the area, starting in the western side of the country and finishing in the east.

## Western Mediterranean

The north coast, to the west of Alexandria, is broken into 4 zones with littoral oolitis sand-dunes, parallel ridges of salt marshes, oolitic limestone, and inland plains. There are few rocky outcrops and few inlets offering harbour facilities. The rainfall, which is mainly in winter, is between 70 and 200 millimetres and has helped to allow 800 species of plants to grow, making up half of the total species for the country.

## Western Desert

The Western Desert makes up 681,000 square kilometres of land which is $^2/_3$ of the total landmass. The main rocks are Eocene limestone with cretaceous (Nubian) sandstone. An igneous uplift in the south west corner has resulted in Gebel Uweinat (1,892 metres) which may have rain every 7 to 10 years, if it is lucky!

The region's largest area is the Eocene limestone plateau (500 metres) which is nearly lifeless, with its shifting sands [the great sand-sheet] which stretch for 500 kilometres from the Libyan border to Farafra. A smaller Selima sand-sheet of miocene limestone is found from the Sudanese border north to Siwa; here the rain may fall once every 10 years, resulting in occasional annual plant-life.

A sheet of Miocene limestone runs from Siwa up to the Qattara Depression where sandy depressions allow some trees and shrubs to grow. The great Eocene plateau has several depressions which were made when water was plentiful, with the action of heat and wind helping to widen them until they reached a water-table. Many of these are now below sea level with Siwa at 30 metres below, and the Qattara Depression at 134 metres below sea level.

These oases make up the main human habitation in the desert today with dates and fruit the main crops. During bird migration, crossing such large areas devoid of water and vegetation, death can result if these oases are not reached in time. The birds may well hit a sandstorm or prevailing winds, be blown off course (from the Nile) or have set off on their journey undernourished. Genuine desert birds have adapted to these harsh conditions by feeding and breeding in very low densities or by being nomadic and moving to new areas of food.

Between 5,000 BC and 2,500 BC, up to 300 millimetres of rain fell in the desert area, resulting in many shrub areas like the Sahal of today including a wide-spread population. Wildlife was also widespread and the ensuring change meant that many species were lost to Egypt, with the human populations moving back to permanent water along the Nile and Delta. Some changes are the direct result of man, but others can be natural changes in weather patterns, as seen along the coast of South America when sea currents change.

With the deserts occupying up to 96% of the landmass at any time, the Nile and its delta offer the main source of fresh water and fertile soils. This resulted in early civilization which changed, especially the mammals, with many 'African' species found this far north. The birds were better adapted to this change and many have been constant for up to 4,000 years as seen on the paintings and hieroglyphs all those years ago.

## The Nile Valley and the Delta

Creating all this was the Blue and White Niles joining north of Khartoum in Sudan, bringing not just water but highly fertile sediments, especially from the Ethiopian Highlands with up to 10 metres of sediment laid down over the years in the valley and the delta. The modern dams have stopped this flow of free fertilisers, resulting in a present-day bill of multi-billion Egyptian pounds. The dam itself will be full of silt in 1,400'years time.

The lack of silt also affects the Delta, especially where it meets the sea, as this area was once a growth of land while now the opposite is happening with erosion washing it away. This is costly to stop. The advantages of the dam are a source of fresh water, cheap electricity and no floods.

The Nile itself, before the dam, ran for 850 kilometres with an average river width of 750 metres and a valley width of up to 10 kilometres. The river flows mainly on the eastern side of the valley with an occasional outcrop of Nubian sandstone from the Eastern Desert. Barrages were first made from 1843 to 1861 to raise the water level to make it easier to irrigate the land. Where the Nile meets the Delta there used to be several distributaries flowing to the sea but these were drawn into two channels, the Rosetta and Damietta, which were the first waterways to have barrages on them.

The total land is 13,000 square kilometres for the Nile Valley but 22,000 square kilometres for the Delta. As the delta meets the sea several saline lakes were made, making up 25% of the Mediterranean wetlands. These areas are constantly under pressure to create new farmland even though fish from these lakes is a major food for the Egyptian people. The Nile and delta are the main production tools for feeding up to 60 million people and exporting rice, fruit and vegetables.

Given the fact that the Nile is a valley running a third of the way into Africa, this offers an obvious choice of a migratory route. Not only does it have water but it has many feeding areas along its way. In recent times since the building of the high dam, there has also been an increase in the growth of phragmites and typha along its route in Egypt. This offers extra feeding and safe roosting sites for many bird species.

The disadvantages come in the lack of high ground for soaring, thus upsetting the raptors, storks, cranes and pelicans. If the food is there then the extra flying/flapping (the burning of energy) can be replaced, so it is interesting to see the majority of the raptors stick to the mountains while the storks, cranes and pelicans can move off the high ground and come down the Nile when the need is there.

As for wintering, the food has to be here and for many ducks it is, but the extreme disturbance from, especially shooting, often means that they move on. But what about 'global warming'? The fact that the ducks no longer come in numbers as they used to do could be related to birds no longer needing to travel so far. This would have to be matched by an increase farther north if the conditions there had become milder. The fact that no northern geese use Egypt now, like they used to do, may also point to this factor. Swans have been found when weather conditions have become more difficult farther north.

The changes in the Delta have resulted in an increase in birds like little gull and whiskered terns, feeding on an increase in insects but there has also been a reduction in aquatic vegetation causing a decline in coots, grebes and diving duck. Where have these birds gone? Are they now using Lake Nasser? And what about these African birds moving north to southern parts of Egypt? Is that also due to Lake Nasser?

## Eastern Desert

This is really a dry mountainous area making up 21% of the landmass, with 223,000 square kilometres. It makes up a backbone stretching from Sudan up to the Cairo-Suez road where the mountains rise to heights of 2,187 metres (Gebel Shayib) through ten other peaks, ranging from 1,437 to 1,751 metres. Across the road the peaks are as low as 140 to 520 metres.

These rugged igneous and metamorphic peaks are some of the highest in Egypt. It is a hyperarid area with little rainfall. What rain does fall is mainly on the east side, leaving a shadow on the west side down to the Nile. The northern half also holds sedimentary rocks with limestone plateaus like Gebel Ataqa, Bahariya, Galala and the Ma'aza plateau, with sandstone found on the Ababda plateau. Farther south a sandstone fringe stretches from the Sudanese border to meet these; here is the largest 'dry' wadi, called Allaqi, in the country with a catchment of 44,000 square kilometres. Some of its lower reaches are now flooded by Lake Nasser.

The Eastern Desert has a long history of mines and quarries with gold, granite uranium and phosphate being mined today. These early miners and settlers destroyed large areas of acacia trees to make charcoal. This tree would have been such a characteristic species in this dry landscape. A number of oil installations are found along the coast bordering the Gulf of Sinai where rigs pipe the oil ashore, making this now a major oil-producing area.

The Gebel Elba area in the south eastern corner is made up of a single igneous outcrop which rises to 1,437 metres. Here the rainfall is slightly higher with the tree and shrub cover being able to catch moisture from the mists from the north easterly winds. This mist is found mainly in the summer months, pushing up the rainfall to as much as 400 millimetres. Due to the good management of the local Bischaran people this landscape has been protected to conserve the game and plants, from which these people make their living. Like some of the bird species, most of this 'plant life' is found nowhere else in Egypt.

This backbone of the Eastern Desert mountains make up the main migration route of millions of raptors each year. They have the heat for soaring and the height for keeping them well up in the atmosphere which allows them to travel farther and faster, with most of them not feeding or drinking on the total migration. They dislike crossing open water and this chain of mountains can take them from central Africa through Egypt, Israel/Jordan, Lebanon, Syria, and Turkey where some go west through the Bosphorus and others east along the Caucasus Mountains. The reverse is the case in the autumn.

## Red Sea Islands

Along the Red Sea are a number of islands made up of igneous core and fossil coral. The living coral around the islands protect them from erosion from the sea and the sea bush, mangrove, also acts as a barrier to the waves. A rarer form of island here is the pure fossil coral island which also has a ring-shaped reef forming a lagoon (atol). Both these forms of islands are homes to many rare sea-birds and turtles of the Western Palearctic.

## Sinai

The Sinai covers 61,000 square kilometres and the two fingers (the Gulfs of Suez and Aqaba) of the Great Rift cut it off from Arabia and most of Egypt, with the Suez Canal finally preventing any land movement of mammals and flightless birds moving from Africa to Asia and vice versa.

The southern peninsula is made up of pre-Cambrian igneous and metamorphic mountains. The north of the range has a cretaceous sandstone face and with its vast temperature and rainfall range the plant-life is very changeable with even alpine plants on the tops. On the Gulf of Suez you can have as little as 13 millimetres of rain while the highest mountains at 2,638 metres on Mount Sinai (Musa), only 50 kilometres away can have 300 millimetres, with some of this falling as snow in winter as temperatures can fall to minus 10 centigrade.

Northern Sinai has more of a Mediterranean feel with saline lagoons and sand-dunes moving inland on to gravel plains with limestone plateaus which consist of cretaceous, eocene and miocene limestone rising up to 1,300 metres. The surrounding gulfs are very different with the Gulf of Aqaba falling down to a depth of 1,800 metres whereas the Gulf of Suez has an average depth of only 25 metres.

The currents in Aqaba offer a wider choice of marine life than the Suez, even though the canal has allowed species from the Mediterranean to mix with those of the Red Sea. The Gulf of Aqaba is also unique in being the most northerly growing area for the mangrove.

The Sinai again offers a migration route for millions of birds. The recent development of the water at Sharm el Sheikh sewage farm has shown that weaker birds will come down to drink when it is on offer and even some, such as white storks, pipits and wagtails, have remained to winter in the area. Local birds have made use of this water and if the water was taken away now, there would be a loss of birds to the area.

The coast of North Sinai is the area with the most new birds colonising the country. Many of these have come over the border from Israel and Gaza and can survive the coast, but not inland. Some have managed to keep along the coast and drop into the Delta and one of these (blackbird) has even gone farther, even into the oases of the Western Desert. With the development of agriculture in North Sinai it will be interesting to see if any more species expand in this direction.

# The Birds of the Pharaohs

**W**hat is surprising about Egypt is that, for so many years, very little information was available about the great bounty of birds using the country at the present time, but a look in the temples and tombs of 4,000 years ago shows that over 70 species were present then, and that most are present today.

**T**oday's birdwatcher cannot fail to be moved by the detail which is seen on these amazing paintings - a time when most people were busy trying just to stay alive. The fact that Baket III at Beni Hassan actually had species of bird not found anywhere else in art or hieroglyphs in Ancient Egypt must have made him a 'birder' in the true sense. The wide variety of species in the acacia trees watching Khnumhotep III clap-netting wildfowl again at Beni Hassan are described by many as the work of a real artist who must have known, so well, the details of the birds.

**B**irds which were every-day features and used in the alphabetical signs of hieroglyphs meant that everybody must have felt for, and known about, these species which, sadly, is not true today. Even birds were gods, like Horus the falcon, and birds even took you into the after-life as, for example, the griffon vulture and its wings. What feeling ran through those ancient Egyptians when the sky was full of those thousands of birds moving through the country on migration? Did birds lead to the first human explorers who wanted to find out where the birds went in winter and returned to in summer?

**M**y first experience of the Nile in 1984 made me want to find all the previously-recorded bird species in those temples and tombs and who would have thought that, when I returned to research this book, I would find new species which had not previously been found. Having found those it must be the case that there are others out there, waiting to be claimed. For example, an otter was found in 1995. This was the first example of this species ever to be recorded in Egypt. That means anyone can take part and add to the list, so that even more of Ancient Egyptian bird-life can be shown to the public at large. This is surely the best advert for modern-day conservation/bio-diversity there will ever be!

**T**o help with your discovery I will try to guide you to the best sites available at the present time. Overall, I am sure the guide will be of great help to you. There exists no better start than the Egyptian Museum in Cairo. Here are many of the classic birds brought from sites such as Meydum with its famous geese. Other examples have, sadly, left the country and can be found in over 20 museums around the world. Some of these are on show, others are left in the vaults for future use. I was aided by Elham Ali with my work in the museum and, if she is not too busy, she may be able to help with your visit.

## The Egyptian Museum, Cairo

## R = ground floor
## P = first floor

With so much history to plough through, a guide can be of great help to find out about many of the aspects of bird-life in Ancient Egypt. These are the ones which I have found and are by no means a full reflection of all the aspects of art and hieroglyphs in the bird world. I have concentrated on the main species, often in every-day life, whereas the truss goose at **R 43** is a relic of days gone by but, as it cannot be identified as a species, then I would tend to leave that one for those interested in cooking! All subjects are at these locations at the time of going to press (1998) but beware of the movement of objects from one area to another - this is why it may be better to ask for Elham!

## Ground Floor

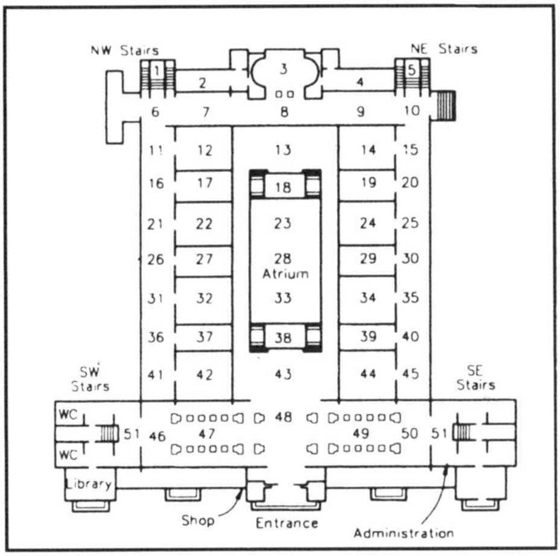


On entering the museum through the main door, in the corner on your left, **R 48**, (JE56001) starts off with the fabulous 'Birds of the Marshes' from King Userkaf, taken from Saqqara, with several species clearly seen, including kingfisher and pied kingfisher, purple gallinule, hoopoe, little bittern and glossy ibis. Is that another bittern? The species lost when the plaster was removed have yet to be identified but it is, nevertheless, a great start to the museum.

**R 41**, found in the corridor, is a 'daily-life' relief from the Mastaba of Nefermaat, Meydum. Here is cormorant, pintail and greylag goose. **R 42** holds a black kite (366) in a wildlife scene where there are also wild dogs and gazelles. **R 36** has a relief of both demoiselle and common crane species from Meydum and **R 32** is where the famous 'Geese of Meydum' can be found, from the Mastaba of Atet with the red breasted, bean and white fronted goose. To the right of that is a splendid Egyptian vulture and quail, also from the Mastaba of Atet. In the same room are hieroglyphic names of birds.

In **R 22** you will find an offering, through a false door, of ducks and geese with wine and bread, while in **R 16** there is a statue with fish and pintail on it. Birds and flowers feature in a relief of Akhenaten in **R 12** while further reliefs are found in **R 8** and **R 3** where teal can be seen, and Horus in flight. The large colourful floor with many pintail present on it, again comes from El Amarna and Akhenaten and is found in **R 28**. There is a great piece of work at **R 10** with Ramese s II as a child with Horus, the falcon god, protecting him.

**A** special mention goes to the world's first bird-bath at **R 14** with lapwing surrounding the sides, and two geese above. It is in fact an offering table with the lapwing representing the foreign people from the Delta/Libya (the main winter range of the lapwing in Egypt!) praising the Pharaoh, in this case, Ramesses II. In **R 49** there are some colourful hieroglyphs of birds while **R 43** has two falcons flying on wooden disks used as toys, with a shoe billed stork on a wooden label close by.

## First floor

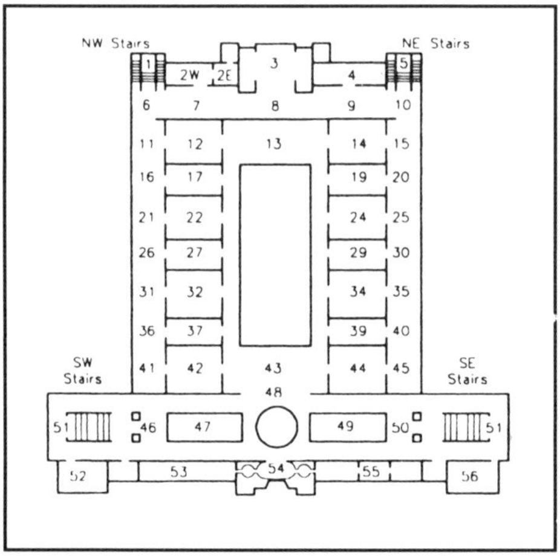


If you enter the first floor by the north west stairs you come to the start of the numbers again, this time with the letter **P** at the beginning. You can start in the Tutankhamun collection where there is a gold sarcophagus at **P 3** which has a lapped faced vulture across the body. Horus appears with Ra and there are several colourful birds in the hieroglyphs around the room. More vultures are found in **P 4**.

A mute swan is found in **P 12** along with a faicon next to it, and a very colourful vulture. Also in the room there is a hunting scene, depicting several duck. The colour continues in **P 17** with some brilliant little egrets on a sarcophagus, along with black kites and Horus.

Several model falcons are found in **P 22** but the problem is that Horus has not been given the name as a particular falcon and several non-descript types of species are often shown. One of these has orange/red legs and undertail, light brown forehead, blue grey back and a pure white breast. The bird must be 95% hobby with another figure showing the same but with a white forehead. In the same room is a male kestrel which, like the hobby, has not been documented before now, probably due to a few errors on the bird. A sacred ibis model is also in the room.

A large wooden model of a mute swan is found in **P 27**. It is $^{2}/_{3}$ life size and great care to detail was made, even including its webbed feet. The swan has recently been given a repair job but still holds its age well, considering it was made between 1991 and 1786 BC.

In contrast, **P 48** contains semi-precious stones along with an ostrich egg. These eggs were often used as vessels or ornaments. The contents of a single ostrich egg could feed as many as 8 people. There is also a mummified falcon and ibis, wrapped in a linen cloth. The sarcophagus of Yuya has a lapped faced vulture and some very colourful hieroglyphs of birds. A feathered pillow is the only one to be found of this same period while the fans are well worth a look. Those good at identifying falcons can try the one on Dirpu which has a very white head, painted wings, blue dark back and pale belly!

More ostrich eggs are found in **P 53** with ovoid vessels made in the shape of ostrich eggs. A large decorative jar contains the intestines/organs of a Horus falcon. There is a falcon's skeleton and a sarcophagus of a sacred ibis also here. There is a future scheme to x-ray Horus's mummies to identify their food remains which may provide, not just new species of birds of prey, but their bird prey as well. Thousands of mummies from Ancient Egypt have still to be looked at.

Lapwing tiles can be seen in **P 44** while the Greco Roman period in **P 39** offers a headless chicken, a cockerel and house sparrows. The throw sticks in **P 34** and **P 25** show how they used to hunt the birds in the swamps of the Delta and make us wonder why we think of boomerangs coming only from Australia! Since the ducks were used as make-up vessels this is supposed to suggest that ducks were erotic symbols but what is wrong with decoys? The 'Book of the Dead' in **P 29** contains several birds including swallows, crag martin, grey heron and a grebe, probably a great crested which has not before been documented.

The reliefs in **P 24** hold a number of birds including red jungle fowl, Egyptian vulture, cranes and falcons, and swallows. There are a number of moulds with great crested grebe, crow and an unusual diver which should not have a crest but also has the wrong feet. There is only one other relief of a diver in the whole of Ancient Egypt found at Karnak. High up is another 'Book of the Dead' while humour comes to the fore with the ink picture of a cat protecting her ducks and eggs from the Ramessid Period.

A wide selection of moulds are found in **P 23** where there are definitely two types of grebes shown. There is no up-turned bill but the bird looks more like black necked grebe (winter visitor) than little grebe which breeds in the country. The great crested grebe is here and is also found on the 'Book of the Dead'. It is mainly a winter visitor but there are some recent records of breeding in the Delta. A cormorant is also here as well as several duck moulds.

Several gold and silver sacred ibises are found in **P 19** along with Horus and vultures and more from Horus on the Pharaoh's religious staff in **P 13**.

These are just some of the examples in the museum and I am sure you will find many more. Other bird reliefs can be found in the Coptic Museum in Cairo with birds in textiles from 300 AD, and also at the Islamic Museum, Cairo where they cannot use human figures so they use plants, animals and birds.

## Saqqara

One of the advantages of Saqqara is that, when the heat becomes overpowering, the hiding places from that exceptional heat are down in the tombs. Here, are a number of sites which offer the enthusiast, not just the odd glimpse of ancient birds, but also the every-day life of the ancient Egyptians. The most famous of these is the Mastaba of Ti. This man was a high court official of the latter part of the fifth dynasty with scenes of work in farming, ship-building, hunting and fishing.

The birds come in the form of the great swamp scene, with several species including both egrets and pied kingfishers with their young. Around the walls are rarities like shoe billed stork, tufted duck and a sandpiper. Turtle doves are frequent as are ducks and geese.

The Mastaba of Mereruka is rated as one of the best examples of art-work anywhere and holds pelican, cormorant, grey heron, greater flamingo and both cranes; and is that a Senegal coucal? Wildlife is widespread here and also includes many gazelles, hippopotami, hyenas, monkeys, oryx, ibex and other antelopes as well as lizards, frogs and locusts. An otter was found here in 1995 which was the first record known to both ancient and modern Egypt. This indicates that new bird species are still possible, even when these fine works of art have been studied for over 100 years!

The Mastaba of Mehu is noted for its ruddy shelduck, white fronted goose, Egyptian goose and many cranes, while Ptahhotep II has the only example of whooper/Bewick swan on its walls, along with the clap-netting of waterfowl, using a decoy/tame heron. Hetepherakhti has bald ibis as well as herons and geese while Neferirtenef has spoonbill and lapwing. Again, this is merely an introduction to finding even more species at this site.

## Beni Hassan

At Beni Hassan upstream from Saqqara lies the home of two of the greatest exhibitions of ancient birds. Baket III (number 15) must be one of the earliest known bird enthusiasts, with his birds set out in habitat preference which is totally different to the normal hunting scenes in most of the other tombs visited. His species, likewise, gave several firsts for birds of Ancient Egypt which were not found anywhere else and he did not add the normally-hunted birds such as turtle dove, ducks and geese. The tomb was, sadly, damaged by being used as a home in more recent times, with religious graffiti written on the walls.

The birds to look out for in the 29 species recorded here include black stork, night heron, purple gallinule, coot, painted snipe, 'ringed' plover, spur winged plover, avocet, roller and pin tailed grouse.

In tomb number 3 Khnumhotep III has a fine portrait of clap-netting with two acacia trees at either end of the net. These trees hold some remarkable birds with no less than 4 masked shrikes in them, a red backed shrike, a hoopoe and two redstarts. Also in the tomb are cormorant, sacred ibis, spoonbill, white fronted goose, wigeon, teal and turtle dove.

## Luxor

Here again we find a large mixture of bird portraits, ranging from a collection at Karnak to birds in general scenes.

Karnak is noted for its botanical gardens at the Festival Hall where there are up to 275 plant species and 52 animal species, of which 38 are birds. The only downward side to these illustrations is that they were supposed to be from a campaign outside Egypt in Western Asia. Most of the bird species have been seen before in art around the other tombs and temples, so where the new species appear, they too could have been records of such species.

One of the new birds is a diver with no obvious up-turned bill, making it either black throated or great northern diver. Great northern is not a common bird from any of these southern latitudes whereas black throated is a rare visitor to the Gulf of Aqaba. Another new bird would be the darter but bone remains have been found at Faiyum and Elphantine Island, Aswan so it could have been one of the ancient birds of Egypt. Great spotted cuckoo is already here breeding and as a migrant, so it should not have been a new bird in Ancient Egypt. Some of the bird species are unidentified, even in family, and need more work on them while species like Egyptian goose, spur winged plover, roller, turtle dove and a sandpiper species can be identified.

Other species to be found while walking around Karnak include several large Egyptian vultures especially on the needles (obelisks), along with glossy and sacred ibis, swallow, lapped face and griffon vultures, pintail, common cranes, lapwing, quail, falcons and a swan. The barn owls in the Festival Hall are very colourful and some seem to have extremely long tails as if they were trying to be harriers! The example of helmeted guinea fowl at Sesostris I Hall is one of the best illustrations of this species found widely in hieroglyphs. It was said to be widespread in the wild up to 200 years ago but was it not domesticated, like in modern times?

## West Bank

Several sites offer good views of these birds and nowhere better than Hatshepsu at Deir al-Bahari where griffon vulture can be found along with common crane, long legged buzzard and even lapwing in flight. The only bee-eater is found here from an expedition to Punt which is thought to be present-day Somalia or Eratria. The pictures here show a scene from this area, including a giraffe and although the bird may have been from this area there are 3 species of bee-eater to choose from in modern Egypt.

A number of tombs have very good bird illustrations, with Horemheb at the Valley of the Kings showing the only record of Dalmatian pelican from Ancient Egypt. The picture shows birds recently captured by clap-netters but they also show eggs so maybe the birds were caught on their nests in the swamplands of the delta region. Ostrich eggs and plumes can be seen in the same tomb, along with the clap-netting of coot and duck.

More birds can be found in the Tombs of the Nobles, with Userhet showing egrets being used as decoys and good hunting scenes, using chariots to chase gazelles. In the Valley of the Queens, the tomb of Nefertari has a good example of a kestrel being shadowed by the phoenix (grey heron).

Upstream from here, the newly-opened 'City of the Vulture Goddess', Nekheb is worth a visit between Esna and Edfu and Horus's temple at Edfu with most birds restricted more to hieroglyphs but it is always worth looking out for them as you never know what you might find, like the Horus on the back of Egypt Air's planes!

## For the record

So far 75 species/families have been recognised in art, of which 63 species are found in hieroglyphs, either as full birds or as parts of them. Of the mummified birds that have been x-rayed (unwrapped!) 44 species have been identified, of which 33 are new and not found in art etc. The modern-day rarities include Abdim's stork and gabar goshawk while migrants such as osprey, honey buzzard, red kite, white tailed eagle, Montagu's harrier, levant and sparrowhawk, spotted, imperial and booted eagle, red footed, saker and peregrine falcon, cuckoo, scops owl and roller were all uncovered.

Residents included black shouldered and black kite, short toed eagle, lesser kestrel, stone curlew, spotted sandgrouse and eagle owl. Wintering birds included hen and pallid harrier, buzzard, merlin, long eared and short eared owl. From the funerary meal of Tutankhamun's tomb came 10 species of bird, of which 3 were new - these were brent goose, gadwell and garganey. Finally, from the archaeological digs at Kom Ombo came a smew, while at Elphantine Island, Aswan there was African fish eagle.

The species which need true identification are the falcons, with Horus causing all sorts of problems. The mummified birds have turned up not only lanner, saker and peregrine but also many other falcons and hawks were used. There are no Barbary falcons yet but with several thousand still to be x-rayed then hopefully some common theme will be found. And what about a hybrid? With so many details often put on Horus, falcons can be cross-bred with other species of falcon, and were the Egyptians the first to do this as they certainly bred many types of birds in captivity?

I asked one of the top falcon breeders in the world this question but he claimed that it was very hard to hybridise the falcons. Mind you, they did build the pyramids, didn't they!

Or is **Horus** just going to be the symbol for all 'Birds of Prey' in ancient and modern times?

## Away from Egypt

With so many artefacts being taken from Egypt and placed around the world in museums and private collections, the birds of Ancient Egypt have travelled too. As many as fifteen museums hold records of birds from Ancient Egypt. The problem is that many museums have varied exhibitions which means that not all of their collection may be on view at any one time.

One of these examples comes from **The British Museum** in London, England where the hunting scene in the papyrus swampland shows a wide variety of colourful bird-life. Decoy herons are being used and birds such as little egret, teal, pied wagtail and Egyptian goose can be seen. The artist seems to have had a cheap buy of blue paint as too much blue coloration has been used in the picture, but is that not the tail of the small bird in the top right hand corner from a bluethroat, a common winter visitor to Egypt? The so-called 'Two Gazelle Palette' shows the only records of fan tailed raven previously called crows! Note the distinctive 'Fantails'. The museum also holds examples of black kite, geese, rock dove, barn owl, quail, swallow, and vultures already identified, and many more may exist.

In the Egyptian Museum in **Berlin**, Germany there is another hunting scene, this time from the Mastaba of Manefer which has 35 birds on it, of which 12 species are of geese and duck. Only 11 species are written about in Houliman's work on ancient birds so are there some new species in this portrait? There is also an interesting scene of quail-catching, as well as other examples of birds, including pelican and pied kingfisher.

The **Metropolitan Museum** of Art in New York, USA has a facsimile of the famous scene from Khnumhotep III's tomb of the clap-netting and the acacia trees. This is now unique as the colours have faded in the original at Beni Hassan. Here also you find griffon vulture, greater flamingo, ostrich, saddle bill stork, black kite and pintail. The **Musée du Louvre** in Paris, France has another papyrus scene with nesting barn owl and is that a dunlin? There is also a scene which shows the catching of golden orioles, along with a hoopoe and kestrel being mobbed by another hoopoe.

**Other museums include** several in USA with the Walkers Museum in Baltimore; the Museum of Fine Arts, Boston; Iowie Museum, Berkley; the Brooklyn Museum, New York; the Oriental Museum, Chicago; and Detroit Museum.

**Other European museums** include Kestner Museum, Hanover, Germany; Royaux d'Art et d'Histoire, Brussels; NY Carlsberg, Copenhagen, Denmark; Museo Egizio, Turin, Italy; and the Leiden Museum, Holland.

Hopefully, this chapter will promote the enjoyment of these **Birds of Ancient Egypt** and allow you to find them in your often limited time spent on holiday or why not visit a museum at home to start your list of these amazing birds? For the **Internet lovers** you can also find many of these birds by travelling through the Egyptian museums while staying at home. Just type **'Egyptology'** in your search section and your journey will begin!

# Nile Cruising

Of all the ways to explore Egypt, the River Nile offers the most luxurious way of spending your time watching the birds. Not only are you transported to the pyramids, temples and tombs but you enjoy the advantage of the height of the cruise boat as you gaze down on wildlife and farming methods, often taking you back those four thousand years.

Today, most cruise boats leave from Luxor and travel to Aswan or conversely, while the 16-day trip from Cairo to Aswan seems to have been forgotten by most people. It is this 600 mile trip from Cairo to Aswan which I have taken part in. There has been a history of cruising in pursuit of bird-life dating back to 1976 when the American Museum of Natural History carried out several winter cruises up to 1979.

There was a short gap to 1981 when the Swan Hellenic Nile Cruises invited the Royal Society for the Protection of Birds (RSPB) from Britain to take part in this spectacle. These were carried out from November 1982 to May 1994 with at least two sailings per year, mainly during March and September but, also other months were covered even in May and June, giving a wider profile of the bird-life than the American winter trips.

Like the American cruise, a day was spent at Lake Qarun reached by going the 90 kilometres by coach across the Western Desert to this special place for birds. The Swan Hellenic Cruise also took the trip to Abu Simbel and, although there was no time set aside for birdwatching, several species were added to the list. Around 120 to 160 birds were often seen on a single cruise.

With so few diversions away from the normal historical trip it is not surprising that many of the temples, pyramids and tombs visited show a remarkable list of species, ranging from the breeding eagle owl to migrating storks which use the height of the buildings to rise and thermal away on migration. This is also not forgetting the permanent birds of Ancient Egypt found on the walls of many of the temples and tombs!

This 16-day cruise takes in the records left by the many birdwatchers in the past and, although migration offers the wider choice of species, I am sure you will be overwhelmed by the sheer numbers and variety on offer. You are travelling with Swan Hellenic from Cairo south, with the trip to Lake Qarun included. The alternatives are mentioned if, for example, you would prefer to stay around Cairo.

## Day 1

For those travelling from London there would be a get-together at the Russell Hotel with a coach to Heathrow where there would be often a buffet lunch and an introduction to the staff. After the flight to Cairo you would join the boat moored on the Nile, often near Maadi, where the darkness of the night would be awake with the calls of the Senegal thick-knees which use the many flat roofs for nesting.

## Day 2 Lake Qarun

Day-break sees you out and about looking for the first birds of the trip. Often these would come in the shape of graceful warblers in the trees along the riverside or of a new subject, the attractive pied kingfisher, using the boat as a perch to dive for fish. Nile cruises have dated the arrival of this bird in Cairo to May 1990, having spread upriver from the south.

Cattle egrets start to fly along the river, not always heading off to find cattle, but more often to the local rubbish tip. Palm doves fly across the river along with hooded crow and the Egyptian race of swallow, with its red underparts, while the first little bittern moves out from the security of the reed bed. Not a bad start but where are those thick-knees? Breakfast at a window seat in the hope of a new bird and then it is off for the coach to take you to the lake.

You soon see why there were so many cattle egrets flying around the river when you pass the Giza Zoo with its colony of 7,000+ pairs, white-washing the parked cars along the street. The first pyramid, surrounded by the many hotels, comes into view but today you drive round them, not even stopping to peer and then it is off along the road into the desert.

This section of road has offered both hoopoe and desert lark, and desert wheatear along with the honking brown necked raven but it is not long before the green of the oasis appears through the front of the coach and it is the crested lark, spur winged plover and more cattle egrets. The first stop allows a look at fan tailed warbler, hoopoe and the Egyptian race of yellow wagtail and then it is off to the water with slender bill gulls calling and waders doing, as most waders do, probe their long/short bill into the mud. Black tailed godwit, avocet, black winged stilt, Temminck's and little stint, marsh sandpiper along with the other sandpipers are present.

Now it is looking for rarities with greater sandplover, Kittlitz's plover, Pacific golden plover and even jack snipe. Amongst the gulls is a single Mediterranean gull, a big black headed gull and a mixture of marsh terns. The small reed beds can hold bittern and little bittern and even little crake has been seen here. Out in the lake are black necked and great crested grebes, shoveler, tufted pochard and ferruginous duck.

Back into the farmland there are specials which may not be seen elsewhere on this trip. The ditches hold painted snipe and Senegal coucal. Passage is in full swing with tawny and red throated pipits in the fields along with black eared, isabelline and northern wheatear. A lesser grey shrike is hunting along the edge of the field while the resident southern great grey shrike perches high on the electric wires. A black shouldered kite hovers in the distance, mobbed by a lanner falcon. Not a bad first full day!

## Day 3 Saqqara and Cruising

The trip to Saqqara by coach offers the chance to look for birds along the agricultural land and the many canals. Pied kingfisher, black shouldered and black kite and little green bee-eater can be seen. The pyramid of Zoser is famous for its breeding eagle owls which can be seen more often on the eastern side of the pyramid. Barn owls breed in the pyramid of Userkaf and eagle owls have also been seen here.

Desert birds are also present here, with both hoopoe lark and desert lark and in winter more records of red tailed wheatear have come from here than anywhere else. Mourning wheatear and blackstart have been recorded here as well as both lanner and sooty falcon. Time has to be spent as much underground here, as a fine collection of ancient birds are located in the tombs, with Mastaba of Ti, Mereruka and Nefer-her-ptah outstanding.

A trip to Memphis has added red avadvant and Senegal coucal but it is soon back to the boat where clamorous reed warbler greet you for an afternoon of cruising. Little egrets join the cattle on your list, with grey heron and osprey soon making themselves known while the large purple gallinules are viewed as the boat glides past them.

## Day 4 Meydum and Cruising

Another coach trip, this time to the step pyramid at Meydum or you can stay on board the boat for the cruise. Over the years, Meydum has produced a mixture of birds ranging from the desert birds such as cream coloured courser, stone curlew and spotted sandgrouse to the farmland birds of red avadavat, red throated, water and tawny pipit, quail, collared dove and southern great grey shrike. Both marsh and Montagu's harrier have been seen, and around the pyramid there have been eagle, little and barn owl, black redstart, blue rock thrush, lesser kestrel and long legged buzzard.

The morning cruising has seen ruddy shelduck, oystercatcher, Senegal coucal and marsh harriers but by the afternoon's sailing you begin to enter into some remarkable birds. The first marbled teal floats by. A painted snipe from the boat! The first pallid harrier is joined by hen harrier and even short eared owl and lapwing are viewed. Can it still be Egypt!

Short toed eagle and displaying long legged buzzards bring you back to earth as does an eagle owl sitting in a tree, watching you as its big head turns as the boat moves by. More raptors include booted and steppe eagle. Sixty five pratincoles is not bad but a sanderling seems well out of its way, as well as red necked grebe. As the light starts to fall, a bittern flies by and a roost of 15 to 20 red footed falcons gives you a smile and you cannot remember moving your feet!

## Day 5 Beni Hassan and Cruising

The first squacco heron is joined by purple heron, even before breakfast. Your figure for pratincoles is shattered by 720 birds in less than an hour. Whimbrel well inland and were they really roseate terns? But today is the day you visit the shrine to a real birdwatcher at No 15, Beni Hassan.

Here we find the tomb of Baket III who comes over as a true birder with over 29 species recorded in his tomb. These were not presented as food as seen in so many other tombs where their existence would be there only for taking into the afterlife for that sole purpose. These were lined out as species, with water birds on one side and land birds on the other. Time has taken its toll on some of the paintings but many can still be recognised.

Another tomb to visit is No 3, of Khnum-hotep III. Here we find birds in an acacia tree with such gems as hoopoe, redstart, and both masked and red backed shrikes. Outside, with the sun beating down, there are still plenty of real birds to spy, with a wide variety having been recorded from here. This is one of the first sites for trumpeter finch often found around the agricultural land near the boat. A flock of 150 were once found here. By the tombs themselves are rock martins with their nests often in the entrances to the tombs. Above the tombs, hooded wheatears can be found.

Walking to the tombs can produce cream coloured courser, desert lark, desert wheatear, spotted sandgrouse and both stone curlew and Senegal thick-knees. The farmland can offer Egyptian nightjar, bluethroat, red throated pipit, quail and southern great grey shrike. Rarities have included red tailed and pied wheatear and two possible female blackstarts and pale phase Eleonora's falcon. Both sooty falcon and booted eagle have been seen here.

Along the river there have been a single mute swan, little gull, lesser crested tern, bittern, gadwell and Temminck's stint. Several parties of cranes have been seen flying over the site. An afternoon stop at Antinopolis can produce the race, schimperi, of the rock dove which is much smaller and purer compared with many of the feral crosses of the common rock dove.

## Day 6 El Til, El Amarna and Cruising

You feel really good about the ancient sites now and another productive area has to be around El Til. In recent years the highlights have been a visit to an area for Egyptian nightjars where a maximum of 15 were counted close to the tombs of El Amarna. Here also has been seen dorcus gazelles, hoopoe larks displaying, and cream coloured courser. The high hills here have attracted migrant raptors such as merlin, peregrine, Barbary and lanner falcons. Lesser spotted and short toed eagles have used the thermals here, as have also long legged buzzard and Egyptian vulture.

Both hooded, pied and mourning wheatears have been found here as well as spectacled warbler, whereas down in the farmland there have been short toed lark, European nightjar, thrush nightingale, olive tree warbler and trumpeter finch. Along the river here have been painted snipe, adult and juvenile pallid harrier, grey wagtail, ferruginous duck, purple heron and spoonbill.

## Day 7 Cruising - All Day

Today is spent cruising along the river which may seem monotonous to some people but, when your boat acts as such an impressive hide, you find that it is very hard to stop concentrating on the river and its surrounds, in case you miss something. This is also when it helps to have a boatload of birdwatchers to avoid missing anything, and so create some impressive counts.

Sand bars attract migrant waders and anything from black tailed godwit to turnstone have been found. The first glossy ibis may appear and number up to 30 for the day, along with increasing numbers of duck including rarities like shelduck. Common and the rare roseate tern have used the sand bars as well as one or two white fronted geese, often mixed in with the increasing number of Egyptian geese. Migrants have included 742 cranes, Dalmatian pelican, lesser grey shrike, booted eagle, lesser kestrel, pallid harriers and, when you reach Gebel Selin, short toed, steppe and spotted eagles have all been seen.

You are very fortunate if you hit a movement of honey buzzards with up to 80 flying back and forth over the boat for a number of hours, along with a short toed eagle. This is when even the non-birder becomes excited as these large birds can be seen without the aid of binoculars. Finally the sunglasses are removed and that good book put down! By the time the number of pied kingfishers reaches staggering proportions (305 between Assiut and Balliana, a distance of 100 kilometres) the non-birders ask, "What is that bird?"

## Day 8 Abydos/Balliana Cruising

A short coach journey brings you to the temples of Seti, Ramesses II and the Osireion. Here you may find blue rock thrush amongst the buildings, along with a variety of migrants passing the site. A number of eagles have been seen from the site with passing steppe, lesser spotted and short toed eagle, while an imperial eagle in December was thought to be wintering.

Both long legged and honey buzzards have been seen over the temples, with Barbary, lanner and sooty falcon also seen, and an impressive 350 cranes circling high. Around the site, short toed lark, trumpeter finch, icterine warbler, black eared and desert wheatear have also been found.

Back on the river, red crested pochard have been seen along with black necked grebes but the highlight of a recent cruise was the 9 yellow billed storks heading south, this being the farthest north this species has been seen on the Nile, with one at Aswan and the remainder at Abu Simbel, this being the main site. Birds of prey have been seen on many of the cruises along this stretch, with several pallid harriers, single spotted and lesser spotted eagles, a single Levant sparrowhawk, wintering booted eagles and sooty falcon.

Greater flamingo make their appearance along with little tern, roller and the farthest-north record of striated heron, with birds being seen from around Nag Hamadi. A special bird of these cruises is the white tailed plover with over 40 being counted from here to Luxor. With Qena past, a regular sighting of crowned sandgrouse comes in view with anything between 2 and 20 birds often coming down to drink on the Nile.

Birds like great spotted cuckoo appear now with breeding recently confirmed from near Aswan where they use hooded crows as hosts. A flock of lesser kestrels was observed from here and white pelicans become more numerous, cutting into the Nile through the Qena gap from the Red Sea. One of the few records of terex sandpiper also comes from this stretch.

Trouble really strikes when you are caught in the swimming pool at the end of the day, with your binoculars nowhere in sight. Was it a lanner or a Barbary falcon which flew by!

## Day 9 Dendera/Hathor and Luxor

An early morning coach ride brings you to the temple of Hathor with a scattering of records which include the first Bonelli's eagle, and rock thrush. Three species of shrike have been seen here with lesser grey being joined by masked and woodchat shrike. The pied wheatear also has black eared, isabelline and mourning wheatear to go with it, while warblers have a singing savi's in the farmland, and Sardinian Warbler as well.

You have had excellent views of a saker falcon which was perched on top of a palm tree. Both honey buzzards and Egyptian vultures have been seen soaring over the temple but you are impressed by the fly-by of 110 greater flamingos by Dendera, heading north as you try to put the other guests on to them. Striated herons, fieldfare and great reed warbler have also been seen on this stretch.

## Day 10 Luxor

The boat now moors in Luxor for two nights, giving plenty of time to explore. With plenty of passage moving through the town you make a point of visiting the Winter Palace Hotel gardens where you find an impressive garden which is one of the best places to find migrating birds due to its maturity and wide variety of trees and shrubs. You hit a good day, starting with masked shrikes, wryneck, bluethroat, chiffchaff, lesser whitethroat, icterine, Sardinian, river and olivaceous warblers and even black kites and steppe buzzards. The most surprising record from here was of a mountain chiffchaff. Local birds are also very good here with up to 4 pairs of Nile Valley sunbirds, hoopoe, bulbul and hooded crow.

Recent visits have also included the Movenpick Hotel on Crocodile Island, 4 kilometres south of Luxor where rufous bush chat, painted snipe, wryneck and white tailed plover have been seen. But with plenty of ancient sites to visit, the birds do not stop coming. On the east bank is Karnak with its great temples and amazing illustrations, many of which are bird illustrations, especially in the Botanical Gardens in the Festival Hall. Even the needles (obelisks) here have some good illustrations of Egyptian vulture.

Up to 2,700 white storks have been seen migrating through Karnak, leaving you even peering through the pillars to count them. Tawny pipit, red footed falcon, hobby, sooty falcon and blue cheeked bee-eaters on the wires by the entrance have all been seen and there is the ever-present barn owl during the light festival, Son et Lumière, to finish off the evening.

## Day 11 Luxor and Cruising to Esna

Across the river is an attractive area of farmland where nightingales, rock thrush, quail and bluethroat have been found. The resthouse at the Ramesseum has a wide variety of birds including lesser kestrel, alpine swift, great spotted cuckoo, lesser grey shrike, icterine and Bonelli's warblers. The mountainous sites of the Valley of the Kings and Queens have had a number of raptors including Barbary, lanner, peregrine and saker falcons, with the lanners having been seen mating.

Migrating cranes, black and white storks, honey buzzards and Egyptian vultures have all been seen soaring over these ridges, while both hooded and mourning wheatears have been seen around the entrances to the tombs. Trumpeter finch is a regular even though birds have been seen escaping the heat during the summer months by hiding in rock crevasses away from the sun!

Leaving Luxor you pass by Crocodile Island and start to hit the largest duck flocks with many more shoveler, wigeon, pintail, teal, tufted duck and ferruginous duck. This is one of the few areas where marbled teal have been seen regularly but, sadly, in small numbers of usually two and three. Ferruginous duck have numbered up to 85 in this area on the cruises but the duck flocks run to several thousand and become very hard to count as they move away before the arrival of the boats.

## Day 12 Esna Edfu and Kom Ombo

After an early walk to the temple of Khnum the boat starts off upriver again on what has been one of the most productive stretches in recent years. One of the most remarkable records from here was of a sacred ibis found during the summer of 1987. The bird was seen by most of the birders on the boat. This would be the first record, since 1892, of a bird so widely used in ancient Egyptian art and in hieroglyphs. A Goliath heron has also made it this far north on the Nile with most records coming from Abu Simbel, and a description of a dove seen along the river here may have been a collared or African collared dove.

The temple for Horus at Edfu has its own owl collection with eagle, little and barn owls all having been seen here as well as breeding kestrels with also some records of lesser kestrel. Back on the river the number of striated herons increase as well as the number of large gulls. Both yellow legged and great black headed have been seen along this stretch. An amazing record of brown noddy, seen well on a sand bar with black headed gulls, did not make the record books but a male namaqua dove did, along with Kittlitz's plovers seen from the boat.

Birds of prey start to appear again with booted, spotted and steppe eagles along with griffon and Egyptian vultures. Pallid harriers increase along with sooty falcons and several sparrowhawks appear along this stretch. Up to 5,100 white storks and over 100 black storks come along with the raptors. Both spotted and crowned sandgrouse have been seen along the river bank coming down to drink, as well as one record of chestnut bellied sandgrouse. Up to 16 purple herons, bittern and 1,038 little egrets have been counted. A crake which was either Ballion's or little was being dwarfed by moorhens, and a pratincole put in an appearance again. In the many duck flocks, up to 43 black necked grebes have been found along with 9 red crested pochard.

## Day 13 Kom Ombo and Final Sailing to Aswan

Kom Ombo has changed a lot since the early cruises and many of the feeding sites for migrants have disappeared, but over the years this was an exceptional stop-over with many firsts for the cruises. The walk to the ptolemaic temple can offer several species including savi's, olivaceous, icterine and Sardinian warblers, bluethroat, red throated pipit, Egyptian nightjar and Spanish sparrow.

The local sewage outflow has offered Kittlitz's plover, greater sandplover, Temminck's stint, white tailed plover and a possible great snipe. The reedy island by the boats has had both bittern and little bittern while the sand bars play host to one of the reliable African skimmer roosts, numbering up to 43 birds.

On the last leg of the cruise, birds still keep on coming with pink backed pelican making its most northerly flight down the Nile, with most records from Abu Simbel. Wintering records of both imperial and spotted eagle come from this stretch and a record count of 5 pallid harriers came from here in the spring. Long legged buzzards and steppe eagles have been seen as well as namaqua dove, black stork and roller.

Approaching Aswan, Montagu's harriers join the many pallid migrating along the Nile, and black kite numbers start to increase. The boat moors with the many other cruise boats but still you have great views across on to the western side where the kites can regularly be seen circling over the mausoleum to the Aga Khan. The boat will be here for another two days during which time you have plenty of opportunity to visit the local sites such as Kitchener's Island where, not only can you find migrants, but breeding little and cattle egrets.

Next to the old Cataract Hotel is a wonderful garden where you can find a variety of bird-life. Over the years, species like rufous bushchat, bluethroat, Sardinian, Bonelli's, wood, savi's, Ruppell's and the rare Menetries warbler have all been found here. Elphantine Island has had olive tree, savi's and Bonelli's warblers along with wryneck, rufous bushchat, water rail, woodchat, red backed and masked shrike while Kitchener's Island has had golden oriole, sub alpine warbler and roller.

Passage can be watched from the boat with large flocks of storks, pelicans, spoonbills, duck, especially garganey and shoveler, and Egyptian geese. Raptors are good value from here with resident lanner and Barbary falcons, passage hobby, peregrine falcon, booted and steppe eagle with vultures including Egyptian, griffon and even lapped faced vulture. Waders both feed and move through the area with records of golden and grey plover, terex sandpiper and black winged pratincoles.

If the option to Abu Simbel is not taken, then there is the possibility of a felucca ride up towards the low dam where you may see striated herons which are common here.

## Day 14 Philae, High Dam and Abu Simbel

The last full day in Upper Egypt can be extremely hot, not just because of the weather but with the sheer effort of trying to cram everything in. Start with Philae, an island named after the original which was flooded as a result of the building of the old dam. The remains were moved to the new island of Agilqiyyah and (fortunately!) re-named Philae.

As this has been one of the regular sites to visit, the number of birds have increased over the years. The regulars are osprey, gull billed tern and whiskered tern, little egret, little green bee-eater, graceful warbler, palm dove and not forgetting the prize for most birders, white crowned black wheatear. These birds can be seen on the rocks around the harbour, on the island and on the adjoining island.

Second to this must be the chance of seeing Egyptian vulture which is often circling to the south of the island, with a number of birds feeding in the village of Shallal on the shores of the lake. A maximum of 30 birds have been seen here as well as the occasional griffon and the rare black vulture.

Migration has swelled the numbers of birds seen from here, with a maximum number of 15,000 white storks and 43 black storks high in the sky, with 30 glossy ibises feeding on the nearby shore. Birds around the island have included red tailed wheatear, thrush nightingale, blue rock thrush, rufous bushchat and even whinchat and blackstart. On the lake has been pink backed pelican and, by the harbour, African collared dove.

A trip to the high dam has given views of white crowned black wheatear, Egyptian vulture and a few migrants including olive tree warbler, pied flycatcher and woodchat shrike. Even the airport at Aswan has offered sightings of eagle owl hunting across the runway!

And finally, it is off to Abu Simbel the site which most birders have heard of before their visit. The word is 'African' and given that so little time is spent here, many have been very lucky. African pied wagtails are a star attraction, feeding around the temple rocks by the lake. They breed out on the islands and fly in to feed along the long shore-line. Namaqua doves have been spotted along with African collared dove and a single yellow billed stork. There was a possible long tailed cormorant, and lapped faced vultures have been seen feeding at the rubbish tip.

Other rarities have included a pair of blackstarts, scrub warbler and Bonelli's eagle over the temple. Other migrants, found often on the walk into the temple grounds, include marsh, great reed, sub-alpine, Ruppell's, orphean and Bonelli's Warblers, Richard's and tawny pipit, masked shrike, pallid harrier and steppe eagle. Larger movements of white storks, cranes and pelicans have been seen, not to mention the white crowned black wheatears.

If you are really lucky and the plane breaks down (on the ground!) and you have to travel back by bus, you stand a good chance of the spotted sandgrouse at the camel feeding area just outside Abu Simbel where up to 50 birds have been recorded. The journey then becomes a hard slog, trying to identify flashes of birds in a fast-moving coach but is still better than missing the experience altogether, and the desert can make up for so much. Some birders did bump into lapped faced vulture on one of these trips. Dream on!

## Day 15 Back to Cairo and The End of a Holiday

Your last morning on the Nile before your flight home and you spend time watching for migrants along the river. But soon it is time to pack and get on to the bus back to the airport for the flight to Cairo. Cairo allows a trip to the museum to find as many birds as possible from ancient times, making sure to see the 'Geese of Meydum' from the Mastaba of Atet and the 'Birds of the Marshes' from the King Userkaf, taken from Saqqara.

There is often an evening trip to the Giza pyramids and the 'Son et Lumière'. Barn owls often appear while eagle owls seem rare here now. Day-time birds have not been as glamorous as at Saqqara but both red tailed and red rumped wheatear have been found here in winter along with peregrine falcon and black redstart. Isabelline wheatear has been seen in spring, along with desert wheatear and desert lark.

The accommodation of the boat has been replaced by a de-luxe hotel with high-rise views over Cairo and, while watching cable TV, your mind floats back to the memories of those great days, birding and enjoying the history of a people who wanted to take birds along with them, even into their after-life.

There are scenes of mind-blowing counts of birds, rarely seen at home, like the 6,752 cattle egrets, 296 little green bee-eaters, 630 pied kingfishers, 298 purple gallinule, 139 purple herons, 95 African skimmers and 21 striated herons. That memory of dawn and of dusk, with the birds coming and going to roost as the boat sails on, temples, pyramids and tombs awash with birds surrounded by the rich agricultural land which feeds those birds from near and far. How can you describe the wonder of all this to your friends? They will surely have to come and experience, first-hand, the Nile for themselves.

N.B. **Swan Hellenic will be running 2 cruises covering the 600 miles in 1998 and hopefully more in the future.**

# Birds of Egypt Check-List

More than 500 different species of birds have been recorded in Egypt or close to the boundaries of the country. There is great scope for adding many new species to the list of this country which makes birdwatching here very exciting, with recent new species coming from out of Africa and also out of Asia. In this list those birds which have been recorded are joined with others that have not but, with so few birdwatchers on this side of the border, many records go amiss.

| | |
|---|---|
| **BR** | Breeding resident |
| **MB** | Breeding migrant coming to the country to breed |
| **EB** | Escaped from captivity and now a breeding resident |
| **FB** | Former breeding bird |
| **WV** | Winter visitor November to February |
| **RWV** | Rare winter visitor |
| **SV** | Non-breeding summer visitor |
| **GM** | General migrant February to May and August to November |
| **SM** | Seen more during the spring migration |
| **AM** | Seen more during the autumn migration |
| **CW** | Coastal wanderer |
| **RCW** | Rare coastal wanderer |
| **AV** | Very rare visitor or seen close to the boundary |
| **HR** | Historical record |

| | | |
|---|---|---|
| Ostrich | Struthio camelus | **BR** |
| Red-throated Diver | Gavia stellata | **RWV** |
| Black-throated Diver | Gavia arctica | **RWV** |
| Little Grebe | Tachybaptus ruficollis | **BR, WV** |
| Great Crested Grebe | Podiceps cristatus | **WV** |
| Red-necked Grebe | Podiceps grisegena | **RWV** |
| Black-necked Grebe | Podiceps nigricollis | **WV** |
| Shy Albatross | Diomedea cauta | **RCW** |
| Schlegel's Petrel | Pterodroma incerta | **RCW** |
| Soft plumaged Petrel | Pterodroma mollis | **RCW** |
| Cory's Shearwater | Calonectris diomedea | **CW** |
| Streaked Shearwater | Calonectris leucomelas | **RCW** |
| Pale footed Shearwater | Puffinus carneipes | **RCW** |
| Sooty Shearwater | Puffinus griseus | **CW** |
| Yelkouan Shearwater | Puffinus yelkouan | **CW** |
| Wedge-tailed Shearwater | Puffinus pacificus | **RCW** |
| Little Shearwater | Puffinus assimilis | **RCW** |
| Persian Shearwater | Puffinus persicus | **RCW** |
| Mascarene Shearwater | Puffinus atrodorsalis | **RCW** |
| Wilson's Storm Petrel | Oceanites oceanicus | **RCW** |
| European Storm Petrel | Hydrobates pelagicus | **RCW** |
| Leach's Storm Petrel | Oceanodroma leucorhoa | **RCW** |
| Maderian Petrel | Oceanodroma castro | **RCW** |
| Red-billed Tropicbird | Phaethon aethereus | **CW, MB** |
| Lesser Frigatebird | Fregata ariel | **RCW** |
| Brown Booby | Sula leucogaster | **CW, MB** |
| Gannet | Sula bassana | **RCW** |
| Cormorant | Phalacrocorax carbo | **WV** |
| Pygmy Cormorant | Phalacrocorax pygmeus | **AV** |

| | | |
|---|---|---|
| Long-tailed Cormorant | Phalacrocrax africanus | **SV, HR** |
| Shag | Phalacrocrax aristotelis | **BR, CV** |
| Darter | Anhinga rufa | **AV** |
| White Pelican | Pelecanus onocrotalus | **GM, WV** |
| Dalmatian Pelican | Pelecanus crispus | **AV** |
| Pink-backed Pelican | Pelecanus rufescens | **SV** |
| Bittern | Botaurus stellaris | **WV** |
| Little Bittern | Ixobrychus minutus | **BR, GM, WV** |
| Night Heron | Nycticorax nycticorax | **BR, GM, WV** |
| Striated Hero | Ardeola striata | **BR, WV** |
| Squacco Heron | Ardeola ralloides | **BR, GM, WV** |
| Cattle Egret | Bubulcus ibis | **BR, GM, WV** |
| Black Heron | Hydranassa ardesiaca | **AV** |
| Western Reef Heron | Egretta gularis | **BR** |
| Little Egret | Egretta garzetta | **BR, GM, WV** |
| Great White Heron | Egretta alba | **GM, WV** |
| Grey Heron | Ardea cinerea | **GM, WV, SV** |
| Black headed Heron | Ardea melanocephala | **AV** |
| Purple Heron | Ardea purpurea | **GM, WV SV** |
| Goliath Heron | Ardea goliath | **BR, AV** |
| Yellow-billed Stork | Mycteria ibis | **SV** |
| Black Stork | Ciconia nigra | **GM, RWV** |
| White Stork | Ciconia ciconia | **GM. WV** |
| Glossy Ibis | Plegadis falcinellus | **GM, WV, BR?** |
| Bald Ibis | Geronticus eremita | **AV?** |
| Sacred Ibis | Threskiornis aethiopicus | **SV** |
| Spoonbill | Platalea leucorodia | **BR, GM, WV** |
| Greater Flamingo | Phoenicopterus ruber | **BR, GM, WV** |
| Lesser Flamingo | Phoenicopterus minor | **AV** |
| Fulvious Whistling Duck | Dendrocygna bicolor | **AV** |
| Mute Swan | Cygnus olor | **RWV** |
| Whooper Swan | Cygnus cygnus | **HR** |
| Bean Goose | Anser fabalis | **RWV** |
| White-fronted Goose | Anser albifrons | **RWV** |
| Lesser White-fronted Goose | Anser erythropus | **AV** |
| Greylag Goose | Anser anser | **HR** |
| Barnacle Goose | Branta leucopsis | **HR** |
| Brent Goose | Branta bernicla | **HR** |
| Red-breasted Goose | Branta ruficollis | **HR** |
| Egyptian Goose | Alopochen aegyptiacus | **BR** |
| Ruddy Shelduck | Tadorna ferruginea | **GM, WV** |
| Shelduck | Tadorna tadorna | **WV** |
| Spur-winged Goose | Plectropterus gambensis | **HR** |
| Cotton Teal | Nettapus coromandelianus | **AV** |
| Wigeon | Anas penelope | **GM, WV** |
| Gadwall | Anas strepera | **GM, WV** |
| Teal | Anas crecca | **GM, WV** |
| Mallard | Anas platyrhynchos | **GM, WV** |
| Pintail | Anas acuta | **WV** |
| Garganey | Anas querquedula | **GM** |
| Shoveler | Anas clypeata | **GM,WV** |
| Marbled Teal | Marmaronetta angustirostris | **RWV** |
| Red-crested Pochard | Netta rufina | **WV** |
| Pochard | Aythya ferina | **GM, WV** |
| Ferruginous Duck | Aythya nyroca | **GM, WV** |
| Tufted Duck | Aythya fuligula | **GM, WV** |

| | | |
|---|---|---|
| Velvet Scoter | Melanitta fusca | **RWV** |
| Smew | Mergus albellus | **HR** |
| Red-breasted Merganser | Mergus serrator | **RWV** |
| White-headed Duck | Oxyura leucocephala | **RWV** |
| Honey Buzzard | Pernis apivoris | **GM** |
| Crested Honey Buzzard | Pernis ptilorhynchus | **AV** |
| Black-shouldered Kite | Elanus caeruleus | **BR** |
| Black Kite | Milvus migrans | **BR, GM, WV** |
| Red Kite | Milvus milvus | **GM** |
| African Fish Eagle | Haliaeetus vocifer | **HR** |
| White-tailed Eagle | Hallaeetus Pelagicus | **AV** |
| Lammergeyer | Gypaetus barbatus | **BR, WV?** |
| Egyptian Vulture | Neophron percnopterus | **BR, GM** |
| Griffon Vulture | Gyps fulvus | **FB, GM, WV** |
| Ruppell's Vulture | Gypus rueppelli | **AV** |
| Lapped faced Vulture | Torgus tracheliotus | **BR, WV?** |
| Black Vulture | Aegypius monachus | **RWV** |
| Short-toed Eagle | Circaetus gallicus | **GM, W** |
| Bateleur | Terathopius ecaudatus | **BR? AV** |
| Marsh Harrier | Circus aeruginosus | **GM, WV** |
| Hen Harrier | Circus cyaneus | **GM, WV** |
| Pallid Harrier | Circus macrourus | **GM, WV** |
| Montagu's Harrier | Cirus pygargus | **GM** |
| Gabar Goshawk | Micronisus gabar | **AV** |
| Goshawk | Accipiter gentilis | **RWV** |
| Sparrowhawk | Accipiter nisus | **GM, WV** |
| Levant Sparrowhawk | Accipiter brevipes | **GM** |
| Buzzard | Buteo buteo | **GM, WV** |
| Steppe Buzzard | Buteo buteo Vulpinus | **GM** |
| Long-legged Buzzard | Buteo rufinus | **BR, GM, WV** |
| Lesser Spotted Eagle | Aquila pomarina | **GM, RWV** |
| Spotted Eagle | Aquila clanga | **GM, RWV** |
| Steppe Eagle | Aquila nipalensis | **GM, RWV** |
| Tawny Eagle | Aquila rapax | **AV, RWV** |
| Imperial Eagle | Aquila heliaca | **GM, RWV** |
| Golden Eagle | Aquila chrysaetos | **BR, RWV** |
| Verreaux's Eagle | Aquila verreauxii | **BR** |
| Booted Eagle | Hieraaetus pennatus | **GM** |
| Bonelli's Eagle | Hieraaetus fasciatus | **BR, GM, RWV** |
| Osprey | Pandion haliaetus | **BR, GM, WV** |
| Kestrel | Falco tinnunculus | **BR, GM, WV** |
| Lesser Kestrel | Falco naumanni | **BR? GM, RWV** |
| Red-footed Falcon | Falco vespertinus | **GM** |
| Merlin | Falco columbarius | **RWV** |
| Hobby | Falco subbuteo | **GM, BM** |
| Eleonora's Falcon | Falco eleonorae | **SM** |
| Sooty Falcon | Falco concolor | **MB** |
| Lanner Falcon | Falco biarmicus | **BR, WV** |
| Saker Falcon | Falco cherrug | **GM, WV** |
| Peregrine Falcon | Falco peregrinus | **GM, WV** |
| Barbary Falcon | Falco pelegrinoides | **BR** |
| Chukar | Alectoris chukar | **BR** |
| Barbary Partridge | Alectoris barbara | **BR?** |
| Sand Partridge | Ammoperdix heyi | **BR** |
| Quail | Coturnix coturnix | **GM, BM?** |
| Water Rail | Rallus aquaticus | **BR, WV** |

Hieroglyphs

Birds in the Delta

Horus + Heiroglyphs

Sacred Ibis

Griffon Vulture

Lapwing

Sandpiper species

Golden Oriole

Pied Kingfisher

Horus at Edfu

Geese of Meydum

Gazelle

Egyptian Goose

Birds of the Marshes

Horus/RamesesII

Lotus flower

Across the Nile to Gezira Island

White Wagtail

Greenfinch

Barn Owl

Tawny Pipit

Hooded Crow

Nightingale

Collared Flycatcher

Woodchat Shrike

Gezira Island-Hoopoe, Collared Flycatcher, Bulbul

Subalpine Warbler

Dahshur-Purple Gallinule, Marsh Harrier, Squacco Heron, Black Kite, Little Egret, Greenshank.

Black shoulded Kite

| | | |
|---|---|---|
| Spotted Crake | Porzana porzana | **WV, GM** |
| Little Crake | Porzana parva | **WV, GM** |
| Baillon's Crake | Porzana pusilla | **WV, GM** |
| Corncrake | Crex crex | **GM** |
| Moorhen | Gallinula chloropus | **BR** |
| Lesser Moorhen | Gallinula angulata | **AV** |
| Purple Gallinule | Porphyrio porphyrio | **BR** |
| Coot | Fulica atra | **WV** |
| Common Crane | Grus grus | **GM** |
| Demoiselle Crane | Anthropoides virgo | **AM** |
| Crowned Crane | Balearica pavonina | **HR** |
| Little Bustard | Tetrax tetrax | **HR** |
| Houbara Bustard | Chlamydotis undulata | **BR, RWV** |
| Great Bustard | Otis tarda | **RWV** |
| Painted Snipe | Rostratula benghalensis | **BR** |
| Osytercatcher | Haematopus ostralegus | **WV** |
| Black-winged Stilt | Himantopus himantopus | **BR, WV** |
| Avocet | Recurvirostra avosetta | **BR, WV** |
| Crab Plover | Dromas ardeola | **SV** |
| Stone Curlew | Burhinus oedicnemus | **BR, WV** |
| Senegal Thick-knee | Burhinus senegalensis | **BR** |
| Egyptian Plover | Pluvianus aegyptius | **HR** |
| Cream-coloured Courser | Cursorius cursor | **BR, GM** |
| Collared Pratincole | Glareola pratincola | **BR, GM** |
| Black-winged Pratincole | Glareola nordmanni | **GM** |
| Oriental Pratincole | Glareola maldivarum | **AV** |
| Little Ringed Plover | Charadrius dubius | **GM, WV, BR** |
| Ringed Plover | Charadrius hiaticula | **GM, WV** |
| Kittlitz's Plover | Charadrius pecuarius | **BR** |
| Kentish Plover | Charadrius alexandrinus | **BR, GM, WV** |
| Lesser Sand Plover | Charadrius mongolus | **SM, AV** |
| Greater Sand Plover | Charadrius leschenaultii | **WV, GM, SV** |
| Caspian Plover | Charadrius asiaticus | **SM** |
| Three banded Plover | Charadrius tricollaris | **AV** |
| Dotterel | Charadrius morinellus | **RWV** |
| Golden Plover | Phuvialis apricaria | **GM, RWV** |
| Pacific Golden Plover | Phuvialis fulva | **AM, AV** |
| Grey Plover | Phuvialis squatarola | **WV, GM** |
| Spur-winged Plover | Hoplopterus spinosus | **BR** |
| Black headed Plover | Hoplopterus tectus | **AV** |
| Red wattled Plover | Hoplopterus indicus | **AV** |
| Sociable Plover | Chettusia gregaria | **RWV** |
| White-tailed Plover | Chettusia leucura | **WV** |
| Lapwing | Vanellus vanellus | **WV** |
| Great Knot | Calidris tenuirostris | **AV** |
| Knot | Calidris canutus | **GM** |
| Sanderling | Calidris alba | **GM, WV** |
| Little Stint | Calidris minuta | **WV, GM** |
| Temminck's Stint | Calidris temminckii | **GM, WV** |
| Long toed Stint | Calidris subminuta | **AV** |
| Pectoral Sandpiper | Calidris melanotos | **AV** |
| Curlew Sandpiper | Calidris ferruginea | **GM** |
| Dunlin | Calidris alpina | **WV, GM** |
| Broad-billed Sandpiper | Limicola falcinellus | **GM** |
| Buff-breasted Sandpiper | Tryngites subruficollis | **HR** |
| Ruff | Philomachus pugnax | **WV, GM** |

| | | |
|---|---|---|
| Jack Snipe | Lymnocryptes minimus | **WV** |
| Common Snipe | Gallinago gallinago | **WV, GM** |
| Great Snipe | Gallinago media | **RWV** |
| Pintail Snipe | Gallinago stenura | **AV** |
| Woodcock | Scolopax rusticola | **RWV** |
| Black-tailed Godwit | Limosa limosa | **GM, WV** |
| Bar-tailed Godwit | Limosa lapponica | **GM, RWV** |
| Whimbrel | Numenius phaeopus | **GM** |
| Slender-billed Curlew | Numenius tenuirostris | **AV, HR?** |
| Curlew | Numenius arquata | **GM, WV** |
| Spotted Redshank | Tringa erythropus | **GM, WV** |
| Redshank | Tringa totanus | **GM, WV** |
| Marsh Sandpiper | Tringa stagnatilis | **GM, WV** |
| Greenshank | Tringa nebularia | **GM, RWV** |
| Green Sandpiper | Tringa ochropus | **GM, WV** |
| Wood Sandpiper | Tringa glareola | **GM, WV** |
| Terek Sandpiper | Xenus cinereus | **GM** |
| Common Sandpiper | Actitis hypoleucus | **GM, WV** |
| Turnstone | Arenaria interpres | **WV** |
| Red-necked Phalarope | Phalaropus lobatus | **GM** |
| Grey Phalarope | Phalaropus fulicarius | **AV** |
| Pomarine Skua | Stercorarius pomarinus | **SM, RWV** |
| Arctic Skua | Stercorarius parasiticus | **SM, RWV** |
| Long-tailed Skua | Stercorarius longicaudus | **AV** |
| Greet Skua | Stercorarius skua | **AV** |
| South Polar Skua | Stercorarius maccormicki | **AV** |
| Sooty Gull | Larus hemprichii | **BR** |
| White-eyed Gull | Larus leucophthalmus | **BR** |
| Great Black-headed Gull | Larus ichthyaetus | **WV, GM** |
| Mediterranean Gull | Larus melanocephalus | **WV** |
| Little Gull | Larus minutus | **WV** |
| Sabine's Gull | Larus sabini | **AV** |
| Black-headed Gull | Larus ridibundus | **WV, GM** |
| Brown headed Gull | Larus brunnicephalus | **AV** |
| Grey headed Gull | Larus cirrocephalus | **AV** |
| Slender-billed Gull | Larus genei | **BR, WV** |
| Audouin's Gull | Larus audouinii | **AM** |
| Common Gull | Larus canus | **WV** |
| Lesser Black-backed Gull | Larus fuscus | **WV, GM** |
| Siberian/Heuglin's Gull | Larus heuglini | **AV** |
| Yellow-legged Gull | Larus cachinnans | **BR, WV** |
| Armenian Gull | Larus armenicus | **WV** |
| Herring Gull | Larus argentatus | **AV** |
| Great black backed Gull | Larus marinus | **WV** |
| Kittiwake | Rissa tridactlya | **GM** |
| Gull-billed Tern | Gelochelidon nilotica | **WV, GM, SV** |
| Caspian Tern | Sterna caspia | **BR, WV** |
| Swift/Crested Tern | Sterna bergii | **MB, WV** |
| Lesser Crested Tern | Sterna bengalensis | **MB, GM** |
| Sandwich Tern | Sterna sandvicensis | **WV, GM** |
| Roseate Tern | Sterna dougallii | **AV** |
| Common Tern | Sterna hirundo | **MB, GM** |
| Arctic Tern | Sterna paradisaea | **GM** |
| White-cheeked Tern | Sterna repressa | **MB, GM** |
| Sooty Tern | Sterna fuscata | **RCV** |
| Common/Brown Noddy | Anous stolidus | **RCV** |

| | | |
|---|---|---|
| Bridled Tern | Sterna anaethetus | **MB, GM** |
| Little Tern | Sterna albifrons | **MB, GM** |
| Saunder's Little Tern | Sterna saundersi | **RCV** |
| Whiskered Tern | Chlidonias hybridus | **GM, WV, SV** |
| Black Tern | Chlidonias niger | **GM, WV** |
| White-winged Black Tern | Chlidonias leucopterus | **GM, WV, SV** |
| African Skimmer | Rynchops flavirostris | **MB, WV** |
| Razorbill | Alca torda | **HR** |
| Lichtenstein's Sandgrouse | Pterocles lichensteinii | **BR** |
| Crowned Sandgrouse | Pterocles coronatus | **BR** |
| Spotted Sandgrouse | Pterocles senegallus | **BR** |
| Chestnut-bellied Sandgrouse | Pterocles exustus | **HV** |
| Black-bellied Sandgrouse | Pterocles orientalis | **WV** |
| Pin-tailed Sandgrouse | Pterocles alchata | **WV** |
| Rock Dove | Columba livia | **BR** |
| Stack Dove | Columba oenas | **WV** |
| Wood Pigeon | Columba palumbus | **RWV** |
| African collared/ Pink-headed Dove | Streptopelia roseogrisea | **BR? GM** |
| Collared Dove | Streptopelia decaocto | **BR** |
| Turtle Dove | Streptopelia turtur | **MB, GM** |
| Rufous Turtle Dove | Streptopelia orientalis | **AV** |
| Palm Dove | Streptepelia senegalensis | **BR** |
| Namaqua Dove | 0ena capensis | **BR, AV** |
| Ring-necked Parakeet | Psittacula krameri | **EB** |
| Alexandrine Parakeet | Psittacula eupatria | **EB** |
| Great Spotted Cuckoo | Clamator glandarius | **GM, BR** |
| Cuckoo | Cuculus canorus | **GM** |
| Senegal Coucal | Centropus senegalensis | **BR** |
| Didric Cuckoo | Chrysococcyx caprius | **AV** |
| Barn Owl | Tyto alba | **BR** |
| Striated Scops Owl | Otus brucei | **RWV** |
| Scops Owl | Otus scops | **GM, BR?** |
| Eagle Owl | Bubo bubo | **BR** |
| Little Owl | Athene noctua | **BR** |
| Hume's Tawny Owl | Strix butleri | **BR** |
| Long-eared Owl | Asio otus | **WV, BR?** |
| Short-eared Owl | Asio fammeus | **WV** |
| Nubian Nightjar | Caprimulgus nubicus | **MB** |
| Nightjar | Caprimulgus europaeus | **GM, AM** |
| Egyptian Nightjar | Caprimulgus aegyptius | **BR, MB** |
| Swift | Apus apus | **GM** |
| Pallid Swift | Apus pallidus | **BR, MB, GM** |
| Alpine Swift | Apus melba | **MB, GM** |
| Little Swift | Apus affinus | **SM** |
| Palm Swift | Cypsiurus parvus | **HR** |
| White-breasted Kingfisher | Halcyon smyrnensis | **BR, GM** |
| Kingfisher | Alcedo atthis | **WV, GM** |
| Pied Kingfisher | Ceryle rudis | **BR** |
| Little Green Bee-eater | Merops orientalis | **BR** |
| Blue-cheeked Bee-eater | Merops supercilliosus | **MB, GM** |
| Bee-eater | Merops apiasterGM, | **MB** |
| Roller | Coracias garrulus | **GM** |
| Abyssinian Roller | Coracias abyssinicus | **AV, SV** |
| Hoopoe | Upupa epops | **MB, GM, WV** |
| Wryneck | Jynx torquilla | **GM** |

| | | |
|---|---|---|
| Syrian Woodpecker | Dendrocopus syriacus | **BR** |
| Black-crowned Finch Lark | Eremopterix nigriceps | **BR, RWV** |
| Chesnut headed Sparrow Lark | Eremopterix signata | **AV** |
| Dunn's Lark | Eremalauda dunni | **BR, WV** |
| Bar-tailed Desert Lark | Ammomanes cincturus | **BR** |
| Desert Lark | Ammomanes deserti | **BR** |
| Hoopoe Lark | Alaemon alaudipes | **BR** |
| Dupont's Lark | Chersophhilus duponti | **BR** |
| Thick-billed Lark | Rhamphocoris clotbey | **MB, RWV** |
| Calandra Lark | Melanocorypha calandra | **WV, GM** |
| Bimaculated Lark | Melanocorypha bimaculata | **GM** |
| Short-toed Lark | Calandrella cinerea | **MB, GM, WV** |
| Hume's short-toed lark | Calandrella acutirostris | **AV** |
| Lesser Short-toed Lark | Calandrella rufescens | **BR, WV** |
| Crested Lark | Galerida cristata | **BR, GM** |
| Thekla Lark | Galerida theklae | **BR** |
| Wood Lark | Lullula aborea | **RWV** |
| Skylark | Alauda arvensis | **WV** |
| Oriental Skylark | Alauda gulgula | **AM** |
| Temminck's Horned Lark | Eremophila bilopa | **BR** |
| Brown throated Sand Martin | Riparia paludicola | **AV** |
| Sand Martin | Riparia riparia | **MB, GM** |
| Banded Martin | Riparia cincta | **AV** |
| Pale Crag Martin | Hirundo obsoleta | **BR** |
| Crag Martin | Hirundo rupestris | **WV** |
| Swallow | Hirundo rustica | **BR, GM** |
| Ethiopian Swallow | Hirundo aethiopica | **AV** |
| Wire tailed Swallow | Hirundo smithii | **AV** |
| Red-rumped Swallow | Hirundo daurica | **GM, SV** |
| House Martin | Delichon urbica | **GM** |
| Richard's Pipit | Anthus richardi | **GM, WV** |
| Blyth's Pipit | Anthus godlewskii | **AV** |
| Tawny Pipit | Anthus campestris | **GM, WV** |
| Long billed Pipit | Anthus similis | **BR, AM?** |
| Olive backed Pipit | Anthus hodgsoni | **AV** |
| Tree Pipit | Anthus trivialis | **GM** |
| Meadow Pipit | Anthus pratensis | **GM, WV** |
| Red throated Pipit | Anthus cervinus | **GM, WV** |
| Water Pipit | Anthus spinoletta | **GM, WV** |
| Buff bellied Pipit | Anthus rubescens japonicus | **RWV** |
| Yellow Wagtail | Motacilla flava | **BR, GM, WV** |
| Citrine Wagtail | Motacilla citreola | **GM** |
| Grey Wagtail | Motacilla cinerea | **WV** |
| White Wagtail | Motacilla alba | **WV** |
| African Pied Wagtail | Motacilla aguimp | **BR** |
| Common Bulbul | Pycnonotus barbatus | **BR** |
| Yellow vented Bulbul | Pycnonotus xanthopygos | **BR** |
| White cheeked Bulbul | Pycnonotus leucogenys | **AV** |
| Grey Hypocolius | Hypocolius ampelinus | **AV** |
| Wren | Troglodytes troglodytes | **RWV** |
| Dunnock | Prunella modularis | **RWV** |
| Rufous Bushchat | Cercotrichas galactotes | **MB GM** |
| Black Bushchat | Cercotrichas podobe | **AV** |
| Robin | Erithacus rubecula | **WV** |
| Thrush Nightingale | Luscinia luscinia | **GM** |
| Nightingale | Luscinia megarhynchos | **GM** |

| | | |
|---|---|---|
| Bluethroat | Luscinia svecica | **WV, GM** |
| Siberian Rubythroat | Luscinia calliope | **AV** |
| White-throated Robin | Irania gutturalis | **AV** |
| Black Redstart | Phoenicurus ochruros | **WV, GM** |
| Redstart | Phoenicurus phoenicurus | **GM, RWV** |
| Blackstart | Cercomela melanura | **BR** |
| Whinchat | Saxicola rubetra | **GM** |
| Stonechat | Saxicola torquata | **WV, GM** |
| Siberian Stonechat | Saxicola torquata maura | **AV** |
| Pied Stonechat | Saxicola caprata | **AV** |
| Isabelline Wheatear | Oenanthe isabellina | **GM, WV, SV** |
| Wheatear | Oenanthe oenanthe | **GM, RWV** |
| Pied Wheatear | Oenanthe pleschanka | **GM, RWV** |
| Cyprus Pied Wheatear | Oenanthe cypriaca | **GM, RWV** |
| Black-eared Wheatear | Oenanthe hispanica | **GM, RWV** |
| Desert Wheatear | Oenanthe deserti | **BR, WV** |
| Finsch's Wheatear | Oenanthe finschii | **RWV** |
| Red-rumped Wheatear | Oenanthe moesta | **BR** |
| Red-tailed Wheatear | Oenanthe xanthoprymna | **RWV** |
| Eastern Pied Wheatear | Oenanthe picata | **AV** |
| Morning Wheatear | Oenanthe lugens | **BR** |
| Hooded Wheatear | Oenanthe monacha | **BR** |
| White-crowned Black Wheatear | Oenanthe leucopyga | **BR** |
| Black Wheatear | Oenanthe leucura | **AV** |
| Rock Thrush | Monticola saxatilis | **GM** |
| Blue Rock Thrush | Monicola solitarius | **WV, GM, BR** |
| Ring Ouzel | Turdus torquatus | **WV** |
| Blackblrd | Turdus merula | **BR, WV** |
| Eye browed Thrush | Turdus obscurus | **AV** |
| Black-throated Thrush | Turdus Ruficollis | **AV** |
| Fieldfare | Turdus pilaris | **RWV** |
| Song Thrush | Turdus philomelos | **WV** |
| Redwing | Turdus iliacus | **RWV** |
| Mistle Thrush | Turdus viscivorus | **RWV** |
| Cetti's Warbler | Cettia cetti | **AV** |
| Fan-tailed Warbler | Cisticola juncidis | **BR** |
| Graceful Warbler | Prinia gracilis | **BR** |
| Scrub Warbler | Scotocerca inquieta | **BR** |
| Pallas's Grasshopper Warbler | Locustella Certhiola | **AV** |
| Grasshopper Warbler | Locustella naevia | **AV** |
| River Warbler | Locustella fluviatilis | **AV** |
| Savi's Warbler | Locusteta lusinioides | **GM, RWV** |
| Aquatic Warbler | Acrocephalus paludicola | **AV** |
| Moustached Warbler | Acrocephalus melanopogon | **GM, RWV** |
| Sedge Warbler | Acrocephalus schoenobaenus | **GM, WV** |
| Paddyfield Warbler | Acrocephalus agricola | **AV** |
| Blyth's Reed Warbler | Acrocephalus dumetorum | **AV** |
| Marsh Warbler | Acrocephalus palustris | **GM** |
| Reed Warbler | Acrocephalus scirpaceus | **MB, GM, RWV** |
| Clamorous Reed Warbler | Acrocephalus stentoreus | **BR** |
| Great Reed Warbler | Acrocephalus arundinaceus | **GM** |
| Barsa Reed Warbler | Acrocephalus griseldis | **AV** |
| Eastern Great Reed Warbler | Acrocephalus orientalis | **AV** |
| Thick billed Warbler | Acrocephalus aedon | **AV** |
| Olivaceous Warbler | Hippolais pallida | **MB, GM, RWV** |
| Booted Warbler | Hippolais caligata | **AV** |

| | | |
|---|---|---|
| Booted 'Syke's' Warbler | Hippolais caligata rama | **AV** |
| Upcher's Warbler | Hippolais languida | **AV** |
| Olive-tree Warbler | Hippolais olivetorum | **GM** |
| Icterine Warbler | Hippolais icterina | **GM** |
| Marmora's Warbler | Sylvia sarda | **HR** |
| Spectacled Warbler | Sylvia conspicillata | **WV, MB** |
| Subalpine Warbler | Sylvia cantillans | **GM, RWV** |
| Menetries Warbler | Sylvia mystacea | **GM, RWV** |
| Sardinian Warbler | Sylvia melanocephala | **GM, BR, WV** |
| Cyprus Warbler | Sylvia melanothorax | **RWV** |
| Ruppell's Warbler | Sylvia rueppellis | **GM** |
| Desert Warbler | Sylvia nana | **RWV** |
| Arabian Warbler | Sylvia leucomelaena | **BR** |
| Orphean Warbler | Sylvia hortensis | **GM** |
| Barred Warbler | Sylvia nisoria | **GM** |
| Lesser Whitethroat | Sylvia curruca | **GM, WV** |
| Whitethroat | Sylvia communis | **GM** |
| Garden Warbler | Sylvia borin | **GM** |
| Blackcap | Sylvia articapilla | **GM, RWV** |
| Green Warbler | Phylloscopus nitidus | **AV** |
| Yellow-browed Warbler | Phylloscopus inornatus | **AV** |
| Hume's Yellow browed Warbler | Phylloscopus humei | **AV** |
| Dusky Warbler | Phylloscopus fuscatus | **AV** |
| Bonelli's Warbler | Phylloscopus bonelli | **GM, RWV** |
| Wood Warbler | Phylloscopus sibilatrix | **GM** |
| Mountain Chiffchaff | Phylloscopus sindianus | **AV** |
| Chiffchaff | Phylloscopus collybita | **GM, WV** |
| Willow Warbler | Phylloscopus trochilus | **GM** |
| Goldcrest | Regulus regulus | **RWV** |
| Firecrest | Regulus ignicapillus | **AV** |
| Spotted Flycatcher | Muscicapa striata | **GM** |
| Red-breasted Flycatcher | Ficedula parva | **AM, GM** |
| Semi-collared Flycatcher | Ficedula semitorquata | **SM, GM** |
| Collared Flycatcher | Ficedula albicollis | **SM, GM** |
| Pied Flycatcher | Muscicapa hypoleuca | **GM** |
| Bearded Tit | Panurus biarmicus | **AV** |
| Arabian Babbler | Turdoides squamiceps | **BR** |
| Fulvous Babbler | Turdoides fulvus | **BR** |
| Great Tit | Parus major | **BR** |
| Penduline Tit | Remiz pendulinus | **RWV** |
| Nile Valley Sunbird | Anthreptes platurus | **BR** |
| Shining Sunbird | Nectarinia habessinica | **BR** |
| Orange-tufted Sunbird | Nectarinia osea | **BR, AM** |
| Golden Oriole | Oriolus oriolus | **GM** |
| Rosy-patched Shrike | Rhodophoneus cruentus | **BR** |
| Isabelline Shrike | Lanius isabellinus | **GM, RWV** |
| Red-becked Shrike | Lanius cristatus | **AM, GM** |
| Lesser Grey Shrike | Lanius schach | **AM, GM** |
| Great Grey Shrike | Lanius exubitor | **BR, GM, WV** |
| Steppe Great Grey Shrike | Lanius exubitor pallidrostris | **AV** |
| Woodchat Shrike | Lanius senator | **GM** |
| Masked Shrike | Lanius nubicus | **SM, GM, BR?** |
| Jackdaw | Corvus monedula | **AV** |
| House Crow | Corvus splendens | **BR** |
| Hooded Crow | Corvus corone | **BR** |
| Rook | Corvus frugilegus | **RWV** |

| | | |
|---|---|---|
| Brown necked Raven | Corvus ruficollis | **BR** |
| Raven | Corvus corvix | **BR, GM?** |
| Fan-tailed Raven | Corvus rhipidurus | **BR** |
| Tristram's Grackle | Onychognathus tristramii | **BR** |
| Starling | Sturnus vulgaris | **WV** |
| Rose-coloured Starling | Sturnus roseus | **AM** |
| House Sparrow | Passer domesticus | **BR** |
| Spanish Sparrow | Passer hispaniolensis | **WV, GM, MB** |
| Dead Sea Sparrow | Passer moabiticus | **AV** |
| Desert Sparrow | Passer simplex | **BR?** |
| Tree Sparrow | Passer montanus | **AV** |
| Sudan Golden Sparrow | Passer luteus | **MB?** |
| Pale Rock Sparrow | Petronia brachydactyla | **AV** |
| Yellow throated Sparrow | Petronia xanthocollis | **AV** |
| Streaked Weaver | Ploceus manyar | **EB** |
| Avadavat | Amandava amandava | **EB** |
| African Silverbill | Euodice cantans | **MB?** |
| Chaffinch | Fringilla coelebs | **RWV** |
| Brambling | Fringilla montifringilla | **RWV** |
| Red-fronted Serin | Serinus pusillus | **AV** |
| Serin | Serinus serinus | **BR, RWV** |
| Syrian Serin | Serinus syriacus | **RWV** |
| Greenfinch | Carduelis chloris | **BR, WV** |
| Goldfinch | Carduelis carduelis | **BR, WV** |
| Siskin | Carduelis spinus | **RWV** |
| Linnet | Carduelis cannabina | **WV** |
| Desert Finch | Rhodospiza obsoleta | **MB, RWV** |
| Trumpeter Finch | Bucanetes githagineus | **BR** |
| Scarlet Rosefinch | Carpodacus erythrinus | **AM** |
| Sinai Rosefinch | Carpodacus synoicus | **BR** |
| Hawfinch | Coccothraustes coccothraustes | **AV** |
| Yellowhammer | Emberiza citrinella | **AV, RWV?** |
| Cirl Bunting | Emberiza cirlus | **HR** |
| Pine Bunting | Emberiza leucocephalos | **AV** |
| House Bunting | Emberiza striolata | **BR** |
| African Rock Bunting | Emberiza tahapisi | **AV** |
| Cinereous Bunting | Emberiza cineracea | **SM, GM** |
| Ortolan Bunting | Emberiza hortulana | **GM** |
| Cretzschmar's Bunting | Emberiza caesia | **GM** |
| Rustic Bunting | Emberiza rustica | **AV** |
| Little Bunting | Emberiza pusilla | **RWV, AM** |
| Yellow-breasted Bunting | Emberiza aureola | **AM** |
| Reed Bunting | Emberiza schoeniculus | **RWV** |
| Red-headed Bunting | Emberiza bruniceps | **AV** |
| Black-headed Bunting | Emberiza melanocephala | **GM** |
| Corn Bunting | Miliaria calandra | **WV** |

## Main Centres for Accommodation and Birdwatching

(Numbers relate to the main map on the back-cover)

| | |
|---|---|
| 1 Cairo | Gezira Island; Giza Zoo and Botanical Gardens; Agricultural Museum; the Pyramids - Giza, Saqqara, Abu Sir, Dahshur and Meydum; Memphis; The Japanese Garden at Helwan; Gebel Asfar; Maadi Petrified Forest; the Nile Barrage; Travelling farther to Lake Qarun; Wadi el Ryan; Wadi el Natrun; Wadi Hagul; $10^{th}$ Ramadan City; Abbassa. |
| 2 Faiyum | Lake Qarun; Wadi el Ryan. |
| 3 Port Said | Lake Manzala; Lake Burullus; Delta; El Malaha. |
| 4 Alexandria | Montaza Gardens; Eastern Harbour and Fort Qait Bay; Lake Maryut; Lake Idku; Wadi el Natrun; El Alamein; Al Alamein; Ras el Hekma. |
| 5 Marsa Matrouh | Ras el Hekma; Al Alamein; Salum; Siwa Oasis; Quattra Depression. |
| 6 El Arish | Lake Bardawil; Rafa; Wadi el Arish. |
| 7 Western Oases | Bahariyya; Farafra; Dakhla and Kharga Oases. |
| 8 Suez | Wadi Hagul; Great Bitter Lakes; Ain Sukhna; Zafarana; the Monastery of St Anthony; the Monastery of St Paul. |
| 9 Hurghada | The Monastery of St Anthony; the Monastery of St Paul; Gebel el Zeit; the islands; the rubbish tip; Safaga-Qena road; Quiser; Bir Beida. |
| 10 Quiser | Bir Beida; Mangrove Bay; Marsa Alam; Shalatein; Gebel Elba. |
| 11 Luxor | West Bank; East Bank; Crocodile Island; Luxor Airport; the river area; Esna; Edfu; Kom Ombo. |
| 12 Aswan | Kom Ombo; Elephantine and Kitchener's Island; Philae; Shallal; the cultivated areas. |
| 13 Abu Simbel | Temples; islands; farmland; camel feeding area. |
| 14 South Sinai | Sharm el Sheikh; sewage works; hotels etc; Ras Mohammed; Nabq; St Catherine's; Wadi Ferian; Taba; Nuweibia; Dahab. |

# Where to Watch Birds

## Cairo

In days gone by, all tourists flying to Egypt would first come to Cairo and then travel from there to wherever they wished to go. This is now not the case, with flights direct to many of the sites such as Luxor, Hurghada or Sharm el Sheikh. This can be good for your current holidays but spare a thought for the capital of history and culture, with either a few days away from your present location, or another holiday!

Like most destinations in Egypt the bird-life is ever-present and even when this bird-life shares it with up to 18 million people. Cairo's birds have adapted to the

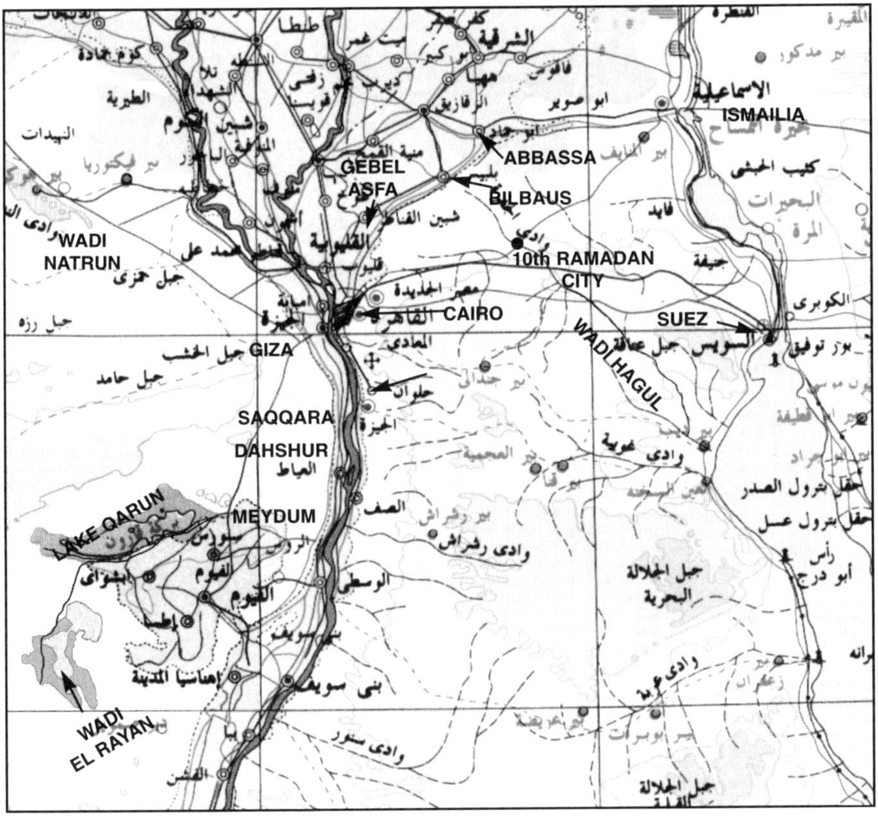

present situation and where some species like the Egyptian and Griffon vultures have moved out, new ones have moved in, for example, the Senegal thick-knees (a bird of cultivated areas) which actually nests on the flat roofs of the city.

The first few minutes out of the airport may offer several new bird species with palm doves flying across the road in front of you, black kites soaring high in the sky, and the Egyptian race of the swallow with its red chest shining in the sun. Many hotels have trees around and it is here that you may find common bulbul along with more European species like hooded crow, house sparrow, blackbird and even hoopoe. If you are really lucky and the trees are in flower there may well be Nile Valley sunbird feeding around the hotel.

The screams of swifts high above the city will be that of pallid swifts and it may be time to go for a walk towards the river. Graceful warblers are found along the river bank, using any type of vegetation, while the white herons flying around are more likely to be cattle egrets which actually breed in the city; a large colony is found at the Giza Zoo where up to 7,200 pairs breed and move around the city for food. They have a far easier time than the traffic below them! Pied kingfishers have moved into the city in recent years and are found at several locations along the river. Wintering birds by the river include black headed gulls, and along the shore are white wagtails. A high-rise hotel room was taken up one year by a birder, and a pair of peregrine falcons were found wintering in the city as well as the first wintering record of sooty falcon. If birds are thin on the ground where you are staying then a trip to Gezira Island is a possibility.

## Gezira Sports Centre

A visit to the Gezira Sports Centre is a way of dropping pace in a speeding city. The centre takes up 140 hectares of the total 251 hectares of the island and only 20 minutes walk from the centre of town. Most people go there to be active but birdwatchers have the ability to wind down and take time to absorb the trees, shrubs and open places which harbour the birds. Many of the previously-mentioned birds are here but it is a sure thing to see them in this environment, especially the sun birds which enjoy, in particular, the lantana camara bushes.

Evening is a good time to see the Senegal thick-knees which leaves the 'high-rises' to come down to feed in the park, along with resident barn owls which use the many mosques for breeding. But it is the migration time that really buzzes here with over 100 species being recorded over the years. You may see species like woodchat shrike, golden oriole and all three species of 'pied' flycatchers. These can sometimes be found at eye level within metres of one another, feeding around the gardens even before the birdwatcher enters the centre.

Warblers can include olivaceous, wood, willow, Bonelli's, blackcap, garden, Sardinian, Ruppell's and whitethroat. The open areas can have tawny, red throated, tree and meadow pipits while hoopoe and wryneck feed around the grounds, with up to 5 species of wheatear recorded, including pied which is a rare find even in the countryside.

Singing migrants can include the lovely voice of nightingale with all three species of bee-eaters seen hunting in the centre; and rare city records of black shouldered kite, pallid harrier, spur winged plover and quail while winter can hold chiffchaffs, bluethroats, black redstart and stonechat with records of long eared owl, robin, song thrush and dunnock. New breeding escapes include both ring necked and Alexandria parakeets with odd records of Fisher's lovebird and budgerigars.

## Agricultural Museum

This is a much smaller area than the sports centre but if you have time, this is well worth a visit combining the history with the present-day birds in the park and especially in the gardens behind. It is found at the end of the 6th October Bridge Road in Mohandsseen! Many of the same species can be seen here with winter birds especially interesting with chiffchaff, bluethroat, robin and song thrush joining the resident sunbirds, little green bee-eaters, hoopoes and even Senegal thick-knees. A sparrowhawk was hunting here on 14th February 1997 and both lesser whitethroats and blackcaps were present.

## Giza Zoo and Botanical Gardens

These two can be combined as they are next-door to one another on the main Giza road, both having a wide selection of exotic trees which are also well worth looking at. The cattle egret colony at the zoo alone is well worth a visit but the migrants also arrive here in good numbers. The sheer number of cats here cannot help the birds but they keep down the mice and rats feeding on the food for the animals and are better than poisons!

Many seats are found around the zoo and the gardens so it is often best to sit and watch for movement in the trees. There is no problem by the cattle egret colony with comings and goings most of the year and even more during breeding season in the spring right through to the autumn. The noise of the traffic cannot compete with the birds and the whole range of breeding, from nest building to fully grown chicks, can often be seen.

## Other city areas

Many parts of the city have a wide range of exotic trees and shrubs which offer shade for the residents in the heat of summer but also act as a magnet for migrating birds. Areas like Garden City are good for passerines while you may be visiting the Citadel and, with its height, be able to see soaring raptors - even blue rock thrush and black redstart have been found here. The tourist site of the pharonic village run by Dr Regab, found on an island in the Nile, has had several records of red avadavat, little bittern and purple gallinule.

## Out of Cairo

Visitors to Cairo always go to view the pyramids and here there is an opportunity to find even more birds.

## Giza Pyramids
[ see also Nile Cruising ]

The first thing you notice when you arrive here is the lack of birds but be patient and you will be rewarded. Over the years several rarer species including red tailed wheatear and peregrine falcon have been found here. Wintering black redstarts have been seen as well as isabelline wheatears. Little, barn and eagle owls have been found with several soaring raptors including lesser kestrel and lanner falcon. If the heat is now getting to you try the resthouse for its shade and local and passing birds while you enjoy a refreshing drink.

## Abu Sir Pyramids

Not yet on most itineraries for ancient monuments but a must for the birder due to the wide range of species. Without the crowds of Giza the birding can be more rewarding, with owls found here as well but far more red foxes! The dunes to the north west have been the best site for Egyptian nightjar and the ride on horse-back from Giza can take in a few more desert species such as desert and hoopoe lark before using the horses to flush the nightjars which blend into the background.

You can cover the same ground on foot but with less success. The nightjars breed here and a number are also here in the winter. There is also a small breeding colony of blue cheeked bee-eaters which use the wires of the local villas as perches. The local orchards and gardens are another magnet for birds, with residents including Senegal thick-knees and collared dove now increasing in this country and good passage migrants like redstarts, Ruppell's, Bonelli's, willow, Sardinian warblers and lesser whitethroats. Northern, black eared and isabelline wheatears use the dunes here as well as short toed lark and skylark.

## Saqqara Pyramids

[ see also Nile Cruising ]

This has been the 'mecca' for viewing eagle owls for some time now with wintering and breeding birds present. They can be found mainly on the eastern side with the nest-site being on the second step down. Up to 2 adults and 2 young can be seen with the young being seen into June. Both barn and little owls are breeding here, with the small pyramid of Userkef next-door holding nest-sites.

This has been the best site for also catching up with hoopoe lark, with up to 2 birds seen around the area, as well as desert lark. Passage has included soaring lesser kestrels, white stork, lanner and sooty falcons. Red tailed wheatear has wintered here several times with the last one being in 1996/97; and both mourning and desert wheatear have been seen here as well as tawny pipit and blue rock thrush.

## Dahshur Pyramids

A recently-opened site which has the potential for some great finds in the future. The greenery is not as close as Abu Sir but it includes a former hunting lake of King Feluke. Here in autumn and winter large concentrations of duck, herons and grebes are found. Up to 6 great white egrets and 4 purple herons were found on 24th October 1997 along with 3 purple Gallinule and 4 ferruginous ducks as well as gadwall, shoveler, teal, wigeon, mallard, pintail, pochard and tufted duck and a late hobby as well as buzzard, sparrowhawk, kestrel and 3 different marsh harriers. The pyramids themselves have all three owls and on the southern side of the White Pyramid is the nest of a brown necked raven. The adults were seen there in 1997 but, with the increase in visitors, may not be there for much longer.

A female desert wheatear was there on 9th February 1997 while a singing and displaying isabelline wheatear was on passage on 30th March 1997. The walk to the lake and marsh can include hoopoe and desert lark while great grey shrikes use the wires on the edge of the desert and passage can include marsh harrier, several wheatears, quail, bee-eaters, sand martins, swifts, greenshank, reed and sedge warblers. Where the marsh meets the farmland there is Senegal coucal, little green bee-eater, fan tailed warbler, clamorous reed warbler, turtle and palm doves. A spotted eagle wintered here in 1997/98.

Back at the Red Pyramid there was a broken water pipe which brought down a number of birds including isabelline wheatear, red throated, tawny and Richard's pipits, subalpine warbler and lesser whitethroats, skylark and Spanish sparrow. There is a small reed bed to the north west of here and several warblers, as well as bluethroats, have been found here.

## Meydum Pyramid

[ see also Nile Cruising ]

The last of the great pyramids with a wealth of agricultural land around, breaking into the desert so quickly. Desert birds have included spotted sandgrouse, cream coloured courser, stone curlew and long legged buzzard while the agriculture has added red avadavat, both Montagu's and marsh harrier, quail, red throated, tawny and water pipits and great grey shrike. Barn owls are present and probably eagle owls.

## Memphis

Most tour groups stop off at this site but there are few bird records from here. Two of the sought-after birds are red avadavat and Senegal coucal, both of which have been found here. An extension of the breeding range of blackbird comes this far south and on both spring and autumn passage there are a number of records of common warblers passing through.

## Helwan Japanese Gardens

These are the only known Japanese gardens found in the Middle East and, although not on the main tourist route, passing migrants do find them and enjoy their foliage. Species like woodchat shrike, wryneck, pied flycatcher and several warblers have been seen here.

## Maadi Petrified Forest

A site dating back 35 million years and worth more than the pyramids put together! Sadly, this amazing historical site is slowly disappearing under an expanding Cairo. The main stack is still worth the visit along with the resident rock martins and mourning wheatears. With millions recently been spent on a golf course in view from here, somehow you think that the world has got its priorities wrong. The nearby Wadi Digla is a good place for wintering spectacled warblers.

## Nile Barrages [Qanater]

When all else has failed a trip is made to Qanater which is 16 kilometres north of Cairo to find 2 of the country's 'Asian' escapes, red avadavat and streaked weaver. This is an easy place to find them and you may also be rewarded with passing migrants as there is a 1 kilometre square area of beautiful gardens which tend to draw the birds. Species like collared flycatcher, golden oriole, woodchat shrike and even greenfinch and serin have been seen here.

## Gebel Asfar

Depending on the traffic this site is about an hour's drive from the centre of the city on the Bilbous road, turning left before the toll gate and following around, keeping left until you pass the industrial area with a track on the left taking you into the Ministry of Agriculture farm. It is the best place to find a number of birds which are very hard elsewhere in the country and on a weekend can be a very pleasant experience. Do not be put off by the fact that the whole area is being fed by the sewage of Cairo. The smell is rarely there and the effect of all this 'green' manure only adds to the sheer number of birds and even mammals!

The growth of the city has still left a number of sand-dunes close to the site and Egyptian nightjar, hoopoe lark and passing wheatears have been found here. The green area soon swallows up the desert and you are fixed on the number of species like hoopoes, black kites, Senegal coucals and thick-knees, spur winged plovers, graceful and fan tailed warblers using the site.

The mixed agriculture has many orchards which even the pipits and wagtails use, even though they are surrounded by trees. Both green and common sandpipers can be found along the many canals as well as the major prize here being the white breasted kingfisher. They are very noisy birds and often can be heard before being seen. They use the water areas as well as the orchards and they have been located in many areas of the farmland.

Painted snipe is also here and can be found in the vegetation of the canals as well as in the settling pools. They are more active at night, so early evening is a good time to look for them but if you flush them they will fly up like jack snipe from near your feet. The bulk of the waders dropping in here are found at the top end of the farm where a number of settling pools can be found.

Over the canal wall a large open area is ear-marked for use as a golf course. This has been very good in the past with large numbers of ruff, wood sandpipers, snipe, lapwing, spur winged, ringed, little ringed and Kentish plovers using the area. Jack snipe have also been found, as well as three banded plover which was the first record for the Western Palearctic.

Both red avadavat and streaked weaver breed here but finding migrants can be difficult with the sheer amount of vegetation available for them. Shrikes can be seen out on the wires while collared flycatcher, subalpine, Ruppell's, Sardinian and olivaceous warblers have all been seen. Birds of prey include black shouldered and black kites, kestrel, sparrowhawk, both common and steppe buzzards. White storks drift through and all three bee-eaters have been seen here. Evening birdwatching has also resulted in views of red fox, jackal and mongoose.

## Faiyum

This area is only 90 kilometres from Cairo and can be easily covered by travelling out of the city or actually staying out in the countryside which will give you added time to enjoy this area. The area has now two main water areas with Lake Qarun the historical area dating back 1.8 million years whereas Wadi el Rayan was formed only in 1973! Lake Qarun has an impressive winter population of waders, gulls and ducks and some years includes flamingos, and over the years the lake has attracted over 200 species of birds.

If you are staying out here it is a wonderful experience to get up in the morning and birdwatch from the lake shore (where often your hotel can be found). The change in the pace of life is the first thing you notice and the endless movement of ducks and gulls across the water as the locals head off to fish on the lake. Winter offers good numbers of both black necked grebe and great crested grebe which often come in close to the shore in the evening and are still there at first light. There was once as many as 10,000 black necked grebes but this has fallen to 3,000 in recent years, with around 1,000 great crested grebes.

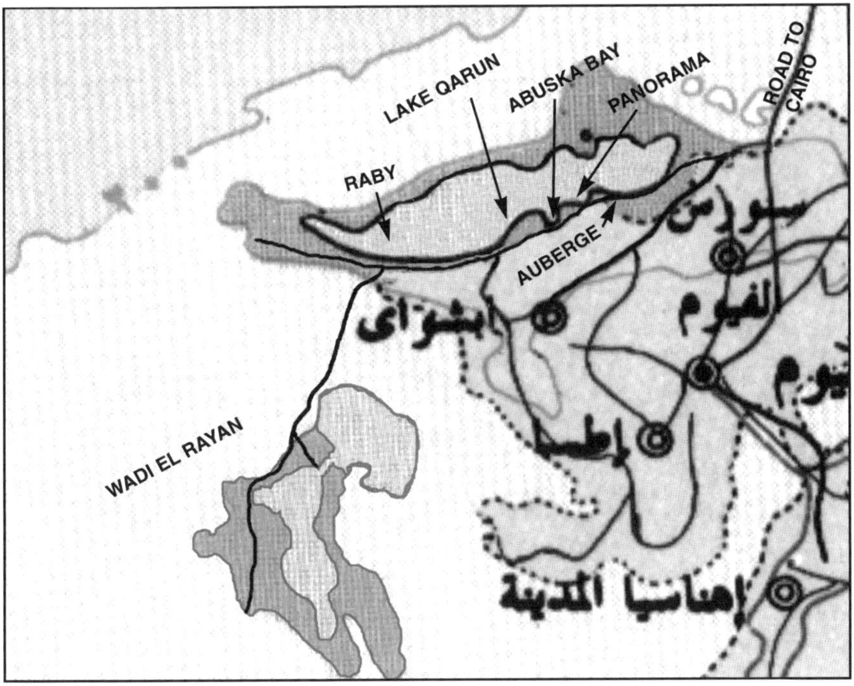


**B**oth black headed and slender bill gulls are present, with a small number of great black headed gulls often reaching 9 in total, with both adults and immatures. They tend to be found on a sand bar to the east of the lake next to the Auberge du Lac Hotel. Other gulls found here include Mediterranean, yellow legged, great black backed, lesser black backed and little gulls. The 'marsh' terns are well represented with both white winged black and whiskered and smaller numbers of black terns. Little terns breed here and large terns like Caspian have been found.

**W**aders come in all sizes, from the wintering little stints to black tailed godwits and avocets. Small numbers of Temminck's actually winter and can be found around the fish ponds in Abuksa Bay. Many other wintering waders are found here including spotted redshank, ruff, wood, marsh and green sandpipers with rare records of red necked phalarope so far inland. The area at the west of the lake near Raby can hold large numbers of waders and duck with greater sandplover here in 1997.

**M**any types of plovers have been found around the lake, with Kentish and spur winged the main breeding ones while Kittlitz's has bred in the past. Grey, golden, ringed, little ringed and greater sand plovers have all been seen while lesser sand, Pacific golden and now sociable and white tailed plover are rarities here. Collared pratincloe breed here along with stone curlew and Senegal thick-knees. Whimbrel pass through with the curlew flocks while dunlin can have curlew sandpipers and even sanderling and turnstone feeding with them.

Painted snipe have been found in many areas of the cultivated land especially at Abuksa Bay where both common and Jack snipe have been found. Senegal coucal is also here and enjoys the heavily-vegetated canal systems running back into the Faiyum. Fewer herons are found here due to the lack of vegetation along the lake but many species pass through, with little bittern breeding along with cattle egrets which use inland trees for nests.

Like the grebes, duck numbers have declined in recent years but shoveler, teal and pintail are here as well as tufted duck and pochard with the odd ferruginous along with them. Migration has many more passerines in the fields by the lake with wheatears, pipits, wagtails and bee-eaters and shrikes using the wires. Passage in the area has seen flocks of white storks, greater flamingos and common cranes while raptors have included harriers, honey buzzards and long legged buzzards.

Wadi el Rayan can seem a quiet place compared with the lake but with no agriculture around the shores, quiet also means that you can be completely on your own which can make the site even more interesting. When the Café Gandlel Wadi, by the waterfall on the southern lake, is not in use you can sit and watch the birds go about their daily chores with clamorous reed warbler, white wagtail, both pied and common kingfisher, little bittern and little egret feeding close to the café.

On the lake there are great crested and black necked grebe, tufted duck, pochard and many coot with a small number of waders using the shore. The increasing area of reed beds allows more herons to use this area and marsh harriers are common in winter along their shores. Even ospreys winter here along with greater black headed gulls.

Small numbers of raptors have been seen passing here in the spring but more watching is desperately needed. Migrants use the reed beds while small isolated bushes in the surrounding desert have attracted species like calandra lark, ortolan bunting and Bonelli's warbler. Desert birds here have included desert wheatear, desert and bar tailed desert lark and hoopoe lark, cream coloured courser and spotted sandgrouse.

## Wadi El Natrun

This is an area 110 kilometres from Cairo and only 90 kilometres from Alexandria on the 'desert' road. The lakes are famous for the salt, sodium carbonate, which was used for the mummification process in Ancient Egypt. Over recent years the area is more famous for one bird, the Kittlitz's plover, which breeds with up to 200 pairs, the largest concentration in the Western Palearctic. The birds are present all year round with a large gathering in the spring, especially on Lake el Fazda.

This is a beautiful area with many quiet parts where you can birdwatch away from the crowds. Most of the area is unspoilt, with agriculture well away from the lakes due to the salts. The springs which bubble up around the lakes often have reed beds growing next to them and here many species of birds can be found. It is one of the few known breeding areas for water rails which can be heard squealing like pigs when you walk on to the marsh areas.

Several waders enjoy these springs with large numbers of ruff and wood sandpipers wintering, as well as Temminck's stint and even great snipe have been found here along with the many common snipe and occasional Jack snipe. Large waders like curlew, redshank and black tailed godwits use the open grasslands while Kentish plover and little stints are found feeding around the lake shores. Smaller

Giza Pyramid with the Sphinx

Saqqara

The flowering Pyramid of Dahshur

The White Pyramid of Dahshur

Birdwatching by the marshes

The farmland around Meydum

Water Hyacynth at Meydum

Eagle Owl

Little Owl

Purple Heron (Immature)

Grey Heron

Bee-eater

Isabelline Wheatear

Ruppell's Warbler

Lake Qarun

Western end of Lake Qarun

Shakshouk village-Lake Quarun

Desert scrub

Wadi el Rayan

Black necked Grebe

Marsh Harrier

Little Crake

Water fall at Wadi el Rayan

Collared Pratincole

Stone Curlew

Fan tailed Warbler

Cream coloured Courser

Wadi El Natrun

Little ringed Plover

Black headed Bunting

Gebel Asfa

White breasted Kingfisher

Delta canal

Oranges and Cattle Egret

Wadi Hagul-Mourning Wheatear, Booted Eagle, Steppe Eagle, Yellow Wagtail (black headed), Ruppell's Warbler, Brown necked Raven, Egyptian Vulture.

Shrikes-Great Grey, Masked (2), Woodchat.

numbers of collared pratincole, cream coloured courser and Egyptian nightjar breed here and painted snipe have also been found.

There is a heavy shooting pressure on this area so few duck species remain but due to the salt, shelduck enjoy the area with up to 500 found in some years. Both black necked and little grebe winter here and there is a small passage of herons with a staging post for common cranes in the desert; this has been damaged by agricultural development and shooting, with several thousand birds using it in spring only.

Black shouldered kites are a new bird to the area and several marsh harriers winter here as well as hen and pallid harriers. Both sparrowhawk and kestrel are found with the occasional merlin, and passage has shown both long legged and honey buzzards with the odd short toed eagles.

The remaining monasteries here may act as good sites for observing migration of passerines but few people stay over and record here. Like so many of these sites each is a green oasis in a desert area and is especially mature. There are no fees for entering and they are well worth a visit even to look for winter birds such as fieldfare, song thrush and dunnock and warblers in the spring and autumn.

Although the area was earmarked for tourist development, at present there are no hotels as such and the only beds available are at the monasteries which are sadly for men only.

## Wadi Hagul

This is another of these sites which can be covered from either of two towns, Cairo or Suez. It lies due south of the main Cairo-Suez road which is 86 kilometres east of Cairo and 47 kilometres west of Suez. The bottom road from Cairo (Maadi) to Suez cuts the wadi road 61 kilometres from Suez and can be used for a round trip from the wadi if you are staying in Cairo. The journey out from Cairo on the main road is on a dual carriageway so can be completed very quickly with brown necked raven, wheatears and larks flashing past. The turning is easily seen with a large sign pointing towards the wadi. Approaching from the north the area seems very flat for several kilometres until the wadi starts taking shape with the first shrubs growing. In 1997 a broken water pipe left many pools close to the road with several red throated pipits, short toed larks, white and yellow wagtails.

Mourning wheatears are found in this first part of the wadi with hoopoe, bar tailed and desert lark likely to be found. Migration can be very good through the wadi with many warblers and pipits using the area. Ruppell's, orphean, Sardinian, Bonelli's, willow warblers, lesser and common whitethroats are regulars, while spectacled warbler bred there in 1995. The best rarity for the area was a wire tailed swallow found by the same group as the spectacled warbler.

Other breeding birds of the area include sooty falcon and scrub warbler which makes the wadi one of the few areas to see these species when based in Cairo. Cream coloured courser is possible along with spotted sandgrouse, brown necked raven and trumpeter finch. Winter birds have included desert warbler and on 27th November 1993, two male Menetries warblers were found along with wintering long legged buzzard but once the raptor passage begins anything is possible.

On 31st March 1997 the first sightings were of distant griffon vultures which had passed over the road and were trying to gain height over Gebel Kahaliya. Not much happened until we reached 3 kilometres before the descent towards the Red Sea. First it was a single short toed eagle and then another and then small groups of 30 to 40 mixed raptors coming due north crossing the road at only 200 metres high. This lasted for 2 hours in an easterly wind with around 800 birds passing until the birds then moved farther west. It would be hard to see these birds cross into Sinai by Suez and most of these must have moved east over the canal and crossed the corner and into Israel.

Steppe buzzards were the commonest bird out of 11 species of raptor, followed by steppe eagles with imperial, spotted, lesser spotted, booted eagles, sparrowhawk, kestrel, black kite, marsh harrier, long legged buzzard, griffon and Egyptian vultures. There were small numbers of white storks with the raptors. The timing for seeing this migration must be very hit and miss in this wadi, but when you do get it right you can see that it can be very rewarding.

## 10th Ramadan City Sewage Works

Like any growing new town there has to be a way of removing the effluent and when the method fits in with wildlife then an attractive wildlife site can be made, even in the desert. It may not be the most desirable tourist site but at the mention of black winged stilt, collared and oriental pratincole, pied and Cyprus pied wheatears, citrine wagtail and many more, those interested in birds will soon be there! The site is only 50 kilometres from Cairo on the Ismailiya road.

## Abbassa

This is found 20 kilometres north east from Bilbeis on the only Bilbeis-Abbassa road running on the left of the Ismailia canal where an old fish farm has been converted to a reed bed right by the road side. Here are thousands of nesting egrets and herons made up of mainly cattle egrets but as well as little egret, night and squacco herons. The road-side view actually looks down on the main nesting area with a second colony found close by and even smaller ones found closer to Abbassa. The birds use the sites for roosting outside the breeding season. The area also holds streaked weaver, avadavat, Senegal coucal and thick-knees and many painted snipe. In the modern fish farms many waders concentrate and even Kittlitz's plover was found there in 1997.

**Accommodation** is widespread around Cairo with excellent hotels in the centre of the city and on the outskirts. It is to be remembered that it can take up to an hour to get out of the city, so planning your accommodation is essential. There are as many as 23 five-star hotels, 15 four-star, 56 three-star and 56 of standard grade accommodation. A small number of floating hotels start from Cairo and go 600 miles to Aswan with Swan Hellenic Cruises and Thomas Cook returning to this tour in 1998. There are also 7 hotels around the Faiyum with two along the shore of Lake Qarun.

Travel is very easy from Cairo with several bus depots, but limited trains with none out towards the Faiyum. Hiring a taxi for the day can be very easy, especially if you ask your hotel to help find a driver who speaks your language. Hire cars are also available along with a driver. Many hotels arrange tours to the ancient sites but none are bird-orientated.

Flights can be arranged to many local sites in the country to Luxor, Aswan, Abu Simbel, Hurghada, Sharm el Sheikh, Marsa Matruh and Alexandria but independent travel is restricted to Luxor along the river, unless you go by Hurghada or via the oases.

## Port Said

**A** town built as a result of the Suez Canal with a population of 400,000, and 212 kilometres from Cairo. It is, in reality, an island built on the edge of Lake Manzala from the waste of the canal, and offers the visitor a wide choice of birdwatching. The port itself is attractive for sea-watching, especially from the harbour mouth which looks out on to the Mediterranean.

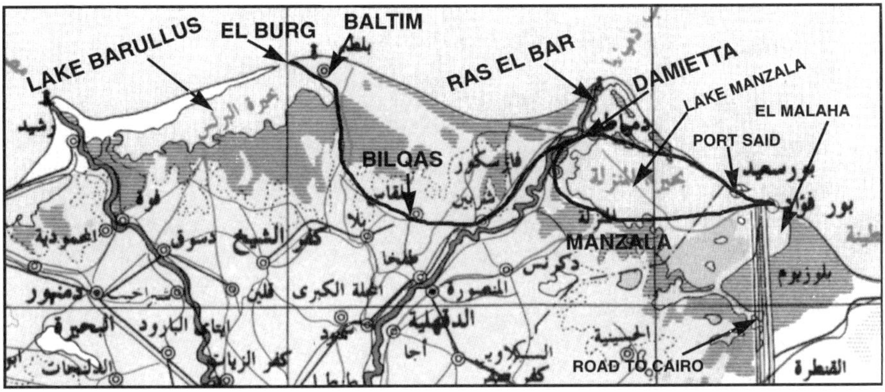


**A** wide variety of gulls have been identified from here with typical winter records including yellow legged, Mediterranean, little, lesser black backed, Siberian and Armenian Gulls. In recent years with better coverage, one or two Audouin's gulls have been found, mainly in the autumn, and there has been an increase in the number of skuas seen with both pomerine and arctic skua found. There is even potential for shearwaters.

**T**he town's gardens, like the Farial Gardens, are attractive for migrants with both pied and collared flycatcher seen in the spring along with wryneck, Spanish sparrow, golden oriole and winter dunnocks. Across the canal on the free ferry to Port Fouad there are a number of rich residential areas where the gardens offer even more in the way of migrants. Local birds, like pied kingfisher, can be found in the harbour in winter.

**L**ake Manzala is right on the doorstep of the town and here you will find a wide assortment of waders, gulls and terns, most of which can be viewed from the road running to Damietta and in the remaining wet areas along the Ismailia road due to the continuing loss of the lake to agricultural land. The highest wintering record of little gulls in the world (47,316) has been counted here and the main wintering flock of whiskered terns (38,741) in the Western Palearctic is also here. Amongst these can be found a number of Armenian, slender billed and black headed gulls. Very small numbers of great black headed gulls as well as sandwich terns winter here. The waders range from those using the Mediterranean shore, such as oystercatcher, turnstone, sanderling and once, in 1990, the crab plover, to the marsh-loving redshank and several sandpipers, including marsh. There is the passage of broad billed sandpipers in April; avocet, little stint, dunlin, plovers including Kittlitz's, greater sand and white tailed plover have all been seen here. Shoveler are the commonest winter duck along with teal, pintail and wigeon.

Breeding birds include several herons, collared pratincole, Kentish and Kittlitz's plovers, painted snipe, little tern and Egyptian nightjar. Black shouldered kites are moving into the area with the expansion of agriculture, and lesser short toed larks are found along the shore. At the western end of the lake is Ras el Bar which holds a number of rocks offshore which are good for gulls and terns with Mediterranean, black headed, little and lesser black backed gulls along with sandwich and common terns. It also looks a good place for sea-watching.

Lake Burullus is by far the most unspoilt of all the delta lakes and sadly is very rarely visited by birdwatchers. There is a picturesque setting looking out on feluccas which carry the many fishermen across the lake from the town of Baltim but, without any commercial offering to the birdwatcher or tourist, the value of this setting seems destined to die like many of the other lakes, due to lack of proper management.

It is not unusual to see tens of marsh harriers floating over the 100 square kilometres of reed bed where the Western Palearctic's highest density of breeding little bitterns, purple gallinules and clamorous reed warblers are found, as well as the lake's breeding collared pratincoles (2,000+ pairs), Egyptian yellow wagtails and lesser short toed larks. There are also wintering ferruginous ducks now fallen from over 6,000 to 500+ but this is still an internationally important number for this declining species.

A number of other wildfowl and waders are found here, all seeming to be in declining numbers especially the duck with shoveler, gadwall, wigeon, teal, pochard, tufted duck and the only record of blue winged teal came from these marshes. Three species of grebes winter here, black necked and great crested with little being the only one breeding here. Several herons are here with great white egret wintering, along with small numbers of greater flamingos. As with Lake Manzala there is a passage of passerines through the area with flycatchers, pipits, larks and warblers likely to be found.

Other parts of the Delta are in easy reach from here and ancient sites like Tanis can be visited along with its resident barn owls and many waders on the agricultural land. There is a large concentration of herons around Abbassa, north east of Bilbeis where regular painted snipe have been found.

El Malaha is found on the east side of the canal next to Port Fouad and over the years has become important for breeding flamingo and slender billed gull. The Plain of Tina which surrounds the lake is sometimes completely flooded right up to the El Arish road which allows grebes and duck to move in, but some years there is no water at all which means the waders dominate with avocet, little stint, dunlin, black tailed godwit and redshank. Along with the flamingos in winter can be over 100 spoonbills and even shelduck and shoveler. Due to the nature of the terrain, few visiting birdwatchers get to see this area.

**Accommodation at Port Said** is varied with only one five-star, one four-star and 8 three-star hotels and several standard grade hotels. You can get here by car, bus, plane or train, and taxis are available to hire thereafter. It is also a 'duty-free' area due to it being next to all those ships bringing goods from around the world, through the Suez Canal.

## Alexandria

This is the largest port in Egypt with a population of 5 million and with a waterfront of 20 kilometres and only 2 kilometres wide. Alexandria is 217 kilometres from Cairo. The city is an ideal place for touring the area with wetlands, desert and sea on your doorstep, including many uncharted areas. The city itself now has a number of local breeding birds which include hoopoe, graceful warbler, pallid swift and the Egyptian race of swallow, as well as the escaped ring necked parakeets.

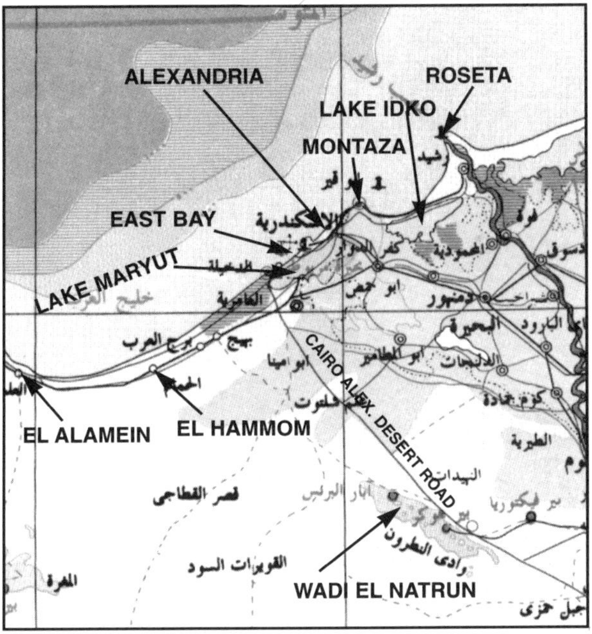


The beautiful and large gardens (140 hectares) of **Montaza Gardens** have been watched for a number of years and are exceptional at turning up wintering species like goldcrest, song thrush, dunnock and siskin. Many of these are hard to find anywhere else in Egypt. Long eared owls are occasionally found here along with wood larks and even hawfinch which is a rare bird in Egypt. Autumn and spring passage is full of warblers, golden orioles, shrikes, whinchats, wheatears, flava wagtails, 'pied' flycatchers with some late records for red breasted flycatcher. The harbour here had 2 Audouin's gulls on 23rd January 1993 with sea-watching a good option on a windy day.

The Eastern harbour, with **Fort Qait** as a viewpoint, makes an ideal site to look for gulls, terns and skuas with records of Audouin's and Heuglin's gulls on the 18th January 1990 and, with a good northerly blow, the potential for shearwaters. **Lake Maryut** is not what it used to be, with only 25% of its former water area remaining as well as a large amount of pollution and noise around but that has never put off the birders! It can be viewed from several roads in the area, especially the desert Cairo road.

The wide reed beds are home to many water rail, clamorous reed warbler, little bittern, purple gallinule with an increasing winter population of greater flamingos and great white egrets to add grace to the site. Black headed and little gulls feed over the water along with whiskered terns while waders and wildfowl pass through the area, with marbled teal and ferruginous ducks both recorded here. Painted snipe, blue cheeked bee-eater, Senegal coucal, red avadavat and streaked weaver can be found here along with many passage birds.

About 30 kilometres to the east of the city is **Lake Idku** which, like much of the Delta area, is under-watched. There are several points along the Alexandria to Rosetta road where you can pull off and view the lake with the best one being just over the bridge from the outflow to the sea. The lake contains a larger number of duck than does Lake Maryut, with wigeon and gadwall enjoying the aquatic vegetation here along with the coot which dive down and pull up strands of weed to help feed the others.

Breeding birds are similar to those of Lake Maryut with little bittern, clamorous reed warbler, water rail, purple gallinule, streaked weaver and Senegal coucal. Above the water are numerous gulls and terns with several thousand whiskered terns on 7th February 1997, along with smaller numbers of little gulls.

The waders can be viewed well from the numerous fish farms along the roadside where avocets are often the commonest wader along with dunlin, little stint, redshank, wood sandpipers and snipe. Rarer finds here include both Jack and great snipe, and also greater sandplovers. Both kingfishers are here in winter with both marsh harrier and black shouldered kites hunting the area. The agricultural land around the area offers many of the local birds such as cattle egrets, spur winged plover, Senegal coucals and crested larks.

Travelling 90 kilometres south from Alexandria a visit to **Wadi el Natrun** is worthwhile with its wide variety of birds, especially the Kittlitz's plovers. A return can be made by the new development from desert to agricultural land which is a wide and open area with few trees established yet. Here on 1st March 1997, by just driving in the area, were large flocks of short toed larks, marsh and pallid harrier, black shouldered kite and long legged buzzard.

Spur winged plovers and crested larks had moved into what was once the territory of cream coloured courser and desert lark. Many wagtails, pipits, Spanish sparrows and wheatears were feeding where once only desert existed. Even pied kingfisher had moved into the new canals with cattle egrets around the new fields, without nesting trees anywhere in sight. Some parts of the road still have desert on one side and green on the other, showing the power of continuous water.

Alexandria is a great base for exploring the Western Mediterranean with the **Marsa Matruh** road ideal for this. The first part of the coast has been taken over by developers with many villas stretching out along the beaches but this often gives access to drive straight to the beach. The many villa gardens are still being cultivated with all sorts of flora which offer the migrants a safe haven when on migration.

Over a million birds are caught each autumn for the pot along the Egyptian Mediterranean. Half of these are quail which are sold alive to markets as far away as Cairo, but along with them are a number of corncrake which is a fast-declining bird in the Western Palearctic due to an increase in modern agricultural methods.

Many birds, like red backed shrike, golden orioles and warblers are eaten by the locals and with no reserves other than Lake Bardawil in North Sinai, this new development may be the only sanctuary the migrants have on their travels north and south across this part of Egypt.

The first real sanctuaries should have been around the war cemeteries of **El Alamein** with thousands of graves of men lost in the battles of World War 2. Here amongst the graves are shrubs, trees and water which are the essential needs of many migrants. In the British cemetery, for example, is a large perimeter of natural vegetation and here migrants gather to feed and should be safe to return to their countries, unlike those gallant men who were not allowed to.

Wintering stonechat are joined by chiffchaff, black redstart and lesser whitethroat; all could be arriving here from places as far afield as Britain, France, Germany and Italy, carrying the memories of home. On 1st March 1997, skylarks were singing here along with feeding meadow pipits all on their way north, and a male Spanish sparrow was sitting by its dome nest - one of the few breeding records for this bird in Egypt. Sadly, gun cartridges were also here indicating that the cemetery is not yet a completely safe place for these birds.

Continuing along the coastal road area there are a number of exotic gardens which hold species like robin, hoopoe, black redstart, chiffchaff and song thrush. The real gem is the **Hotel Al Alamein near Sidi Abd el Rhman** which is also sometimes referred to as the name of the village. This hotel which is over 30 years old shows what can be done to welcome birds on this coastline. Over the years, a wide range of species have been found in this maturing vegetation which ranges from large palms to reeds and rushes around the wetlands.

The hotel is empty for the winter and does not open until May, leaving a very quiet time for the birds. Rock thrushes can be joined by fieldfare while cetti's warbler have been seen with moustached warblers which are thought to winter here. The wagtails are joined by meadow pipits and skylarks while the wheatears join the many black redstarts. Lesser grey shrike, savi's warbler, red breasted flycatcher and many more make this a site well worth visiting.

The marshes next-door held a wide mixture of waders on 28th February 1997 with green, wood, common and marsh sandpipers and several redshank, a pair of ruff, ringed, little ringed and Kentish plovers. There were several bluethroats, a moorhen and a little crake feeding out in the open and all these were watched from the hotel after a redshank had called. bringing the wetland to my attention.

The many fig orchards are a characteristic feature of this area and migrants make their way to them during passage but the bird everyone is looking for along this coast is the red rumped wheatear. Once they were fairly common along here but in 1994 only one site was known to hold a pair of these wonderful birds. This was at the **El Omayed protected area near El Hammam**. This is a reserve covering 5,000 hectares and is 83 kilometres from Alexandria. It was set up to show the prominent ridges running parallel with the coast which indicates the shifting position of the coast itself. Several desert species breed here including hoopoe lark, desert wheatear and cream coloured courser.

In the open areas along this coast there is a European feel with winter migrants including starling, linnet, stonechat, meadow pipit and skylark with fewer records of wood and calandra lark, corn bunting and chaffinch.

A promontory well worth looking at is **Ras el Hekma**. As well as these wintering birds the sea here offers the chance to see species like shag which breeds on a nearby island. Sea-watching is possible (but not on the very point which is privately owned) with shearwaters, gulls and petrels a possibility, with one record of a long tailed skua from here on 2nd October 1994 and Sabine's gull seen close by.

**Accommodation in Alexandria** is very easy to find with up to 3 five-star, 8 four-star and 13 three-star hotels as well as many other standard grade hotels. Due to the length of the town a good bet is to find a hotel near the Montaza Gardens if it is during migration or winter time, whereas if you fancy sea-watching try the eastern harbour area. Alexandria can be reached by plane, train, bus or car and taxis are good for travelling around the area.

El Alamein has four hotels but you need to check that they are all open, especially in the winter months. The new development along this coast may also build new accommodation for foreign tourists, so changes are bound to happen.

## Marsa Matrouh

This is a seaside town, with a population of 80,000, famous as the headquarters of Rommel during the El Alamein campaign. It is 298 kilometres from Alexandria and only 226 kilometres from the Libyan border. This makes it look as if it is well out on a limb but it is an area which should be given more time by the birder.

A former ringing study at Bahig, east of here, from 1966 to 1973 totalled 76,000 birds of over 75 species so this wide variety could easily be found anywhere along this coastline with towns especially, drawing migrants due to their amazing power of light. The gardens along the Corniche offer a good chance of species with winter robins, song thrush, chiffchaff, stonechat, black redstart, meadow pipits and linnets, all using the town. Warblers, orioles, shrikes and pipits move through during spring and autumn with some raptors even making the long crossing of the Mediterranean with hen and marsh harrier, honey buzzard and hobby. Several lakes are found along the headland with many of them being developed around tourist accommodation. There are few records from here with duck such as garganey and waders generally scattered around them.

The star birds for this area are often resident, with Dupont's and thick billed larks both having been seen in recent years. The main area for the Dupont's lark is on the Marsa to Salum road and 52 kilometres south on the Siwa road. The farthest east they have been seen is 40 kilometres which leaves a lot of land away from roads not yet explored. The thick billed lark was found only along the Siwa road again at the point 52 kilometres away from Marsa. Here 15 birds were found in 1994, with breeding confirmed in 1995. Both Temminck's horned lark and bar tailed desert lark are found here if you fail on the other two.

Travelling west to Salum is found the Thekla lark which has only just been re-found after a gap of nearly 70 years. The birds were thought to re-colonise from the Libyan side of the border, with several males singing in recent years. The bird which is very hard to find here is the Barbary partridge which may well have been hunted to extinction along with the houbara bustard which used to be common here. The other bird found breeding, only along this coast, is the raven which again has crept over the border from Libya.

Marsa Matruh is a great stop-over for travelling down to **Siwa Oasis**. This is a trip back in time of 300 kilometres. The original Berber settlers were drawn here by the island of green surrounded by the desert which they now boast about, having 300,000 palm trees and 70,000 olive trees. As you can imagine little bird work has been done this far inland but over 50 species have been identified over the years.

Local breeding birds include palm dove, rufus bush chat, white crowned black wheatear, eagle owl, lanner falcon and a new bird is the blackbird with 10 pairs in 1994. The desert areas hold both hoopoe and Temminck's horned lark. The only open water comes in the form of up to 18 lakes with 150 artesian wells feeding them. Up to 70 greater flamingo have been counted, feeding in them with a number of herons finding their way here including great white egret and grey heron, along with glossy ibis, white stork, common crane, gull billed tern, whiskered and white winged black tern and coot along with several waders around their edges.

Wintering blue rock thrush can be seen along with stonechat, desert wheatear, and chiffchaffs, while passage has had bee-eater, hoopoe, sand martin, redstart, nightingale, subalpine, Ruppell's and Sardinian warblers. The nearby Qattara Depression is the largest area below sea level in Egypt with 19,500 square kilometres. The lowest point is 134 metres below sea level. There is little access here and again, little bird information, but it is still thought that cheetah survives here in ever-decreasing numbers.

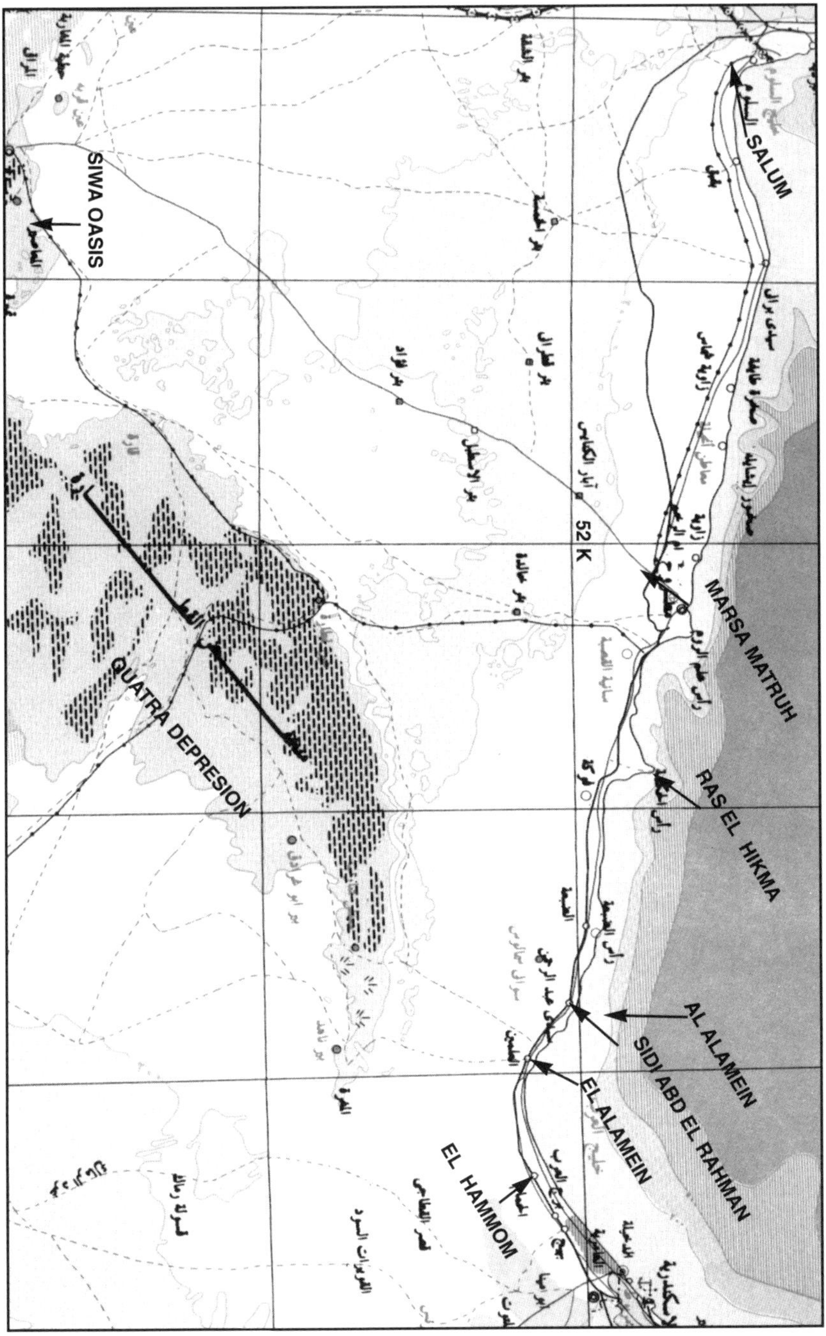
SALUM
SIWA OASIS
52 K
MARSA MATRUH
RAS EL HIKMA
QUATRA DEPRESSION
AL ALAMEIN
SIDI ABD EL RAHMAN
EL ALAMEIN
EL HAMMOM

**Accommodation** is mainly in Marsa Matruh with 12 three-star hotels and 14 standard grade. Development is widely spread around here with new hotels also a possibility. Siwa has only one or two hotels none of which is of a high standard due to the lack of tourists but new development may well be only around the corner.

You can get to Matruh by car, bus, train or plane but only by car or bus to Siwa. Bicycles are a good way to travel around Siwa due to the lack of motorised traffic with many water tanks for you to cool off in.

## El Arish

This town is centred on the North Sinai coast with the Mediterranean rolling onto its beaches. It has a population of 40,000 and is 325 kilometres from Cairo and only 41 kilometres from the Israeli and Palestinian border at Rafa.

It is another summer resort for the Egyptians with several developments for tourists. It acts as a filter for new breeding species moving out from Israel, with recent records of Syrian woodpecker, spotted flycatcher, bee-eater, the European race of the swallow and great tit. It is also in the line of mass migration moving down in the autumn through Israel, with the town holding many migrants.

If you prefer to do your birdwatching on a reserve then the best site to witness this movement is in the protected area of Zaraniq which is an impressive area of salt marsh, salt pans, mud flats, marine open water and sand dunes covering 70,000 hectares, an area of 230 square kilometres which is part of the area better known as Lake Bardawil. It is only 35 kilometres from El Arish and over 230 species of birds have been seen here, with only 8 species regularly breeding.

The breeding birds include internationally large numbers of little tern (1,200 pairs) and Kentish plover (1,900 pairs). Other birds include palm dove, great grey shrike, hoopoe and crested lark and spur winged plover. This is a new site for breeding avocets, and summering flamingos have been tempted to breed by artificial nest-sites being built but as yet with no luck. There is also a small summering flock of greater sandplover.

The main use for the birds then is during migration and winter, with autumn by far the best time to witness this migration. Large numbers of herons, especially little bittern, night, squacco and purple heron pass by with up to 2,000 a day recorded along with white pelicans and both whiskered and white winged black terns. Garganey used to be counted in the 100,000s but now sadly, it is well below this number with day totals more like 22,000. Other declines are in the number of avocets and small waders.

The total variety of waders is still strong with 44 species recorded, including terex and broad billed sandpiper, white tailed and Pacific golden plover and both grey and red necked phalarope. This site is competing with the rubbish tip at Hurghada as the best place to find black winged pratincole.

The passerines fall in the 1,000s and are found in spring nearer the observatory while in autumn they are more likely to be closer to the shore. Passage can start as early as July with the first lesser whitethroats and chiffchaffs followed by lesser grey shrike, black headed buntings and olive tree warblers with September showing full movement with warblers like Ruppell's, orphean, barred, icterine, Bonelli's warbler and whitethroat, red backed shrikes, ortolan and Cretzschmar's adding to the short toed larks, red throated pipits, black eared and isabelline wheatears.

Late movement of red breasted flycatchers sometimes coincides with pied and Cyprus pied wheatears, with odd rarities of yellow breasted bunting and citrine wagtail. Raptors are better in the spring but both Montagu's and pallid harriers pass through, along with honey buzzards. The spring movement can be spectacular with the year 1990 being outstanding when 110,000 birds of prey (24 species) moved through with 6,000 white pelicans. The regulars are lesser spotted, short toed and steppe eagles with smaller numbers of falcons.

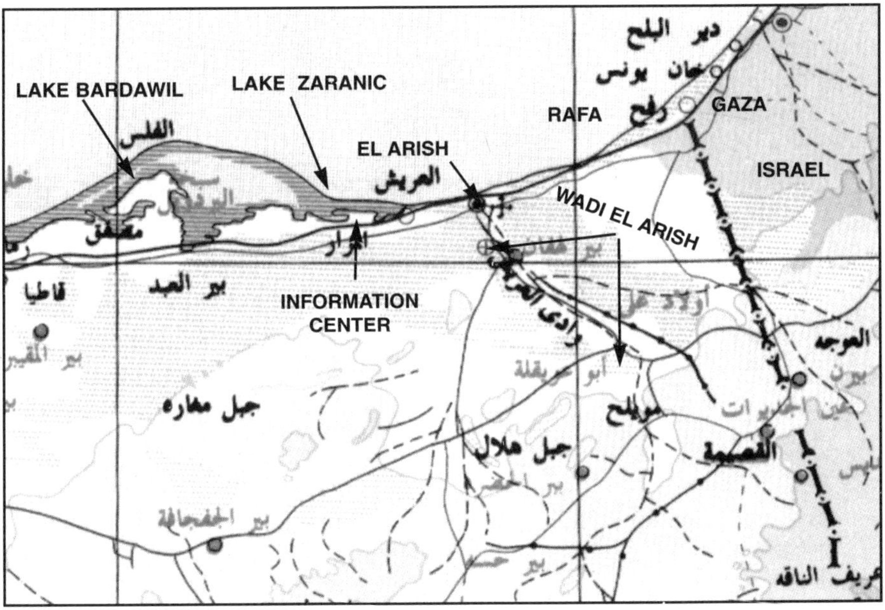


Wintering birds include up to 10,000 flamingos and a new addition is 20,000 cormorants which have come out of Russia. This is a worrying problem for the reserve which has an annual fish catch of 1/4 million tons, and local fisherman are already shooting the birds outside the reserve. Sea-watching can be good here with skuas, gulls, petrels and shearwaters. Armenian gulls are now a regular wintering bird while a small number of Audouin's gulls are found here in the autumn.

There is also a wide variety of other wildlife with lesser gerbils, long eared hedgehogs, fennec fox and up to 3 species of turtle on the beach, with 2 of them breeding. Up to 16 species of reptiles have been recorded here as well as on the sand-dunes which offer more in the way of flora. The reserve has its own self-catering accommodation which sleeps up to 6 people if you want to wake up on the spot.

To the east of El Arish is **Rafa** which is a border town not only with Israel but with Gaza. Again, like El Arish, new breeding birds can pass over the border to breed and the white breasted kingfisher is one of these. A single pair is found here with a long distance before the next pairs are found in the Delta and especially Gebel Asfar near Cairo. The yellow vented bulbul is one which has stuck to the Sinai and is found west as far as El Arish but south-moving along the new hotels on the Gulf of Aqaba. Also here are serin, hobby, chukar Arabian babbler and Sardinian warbler.

The surrounding area has many roads but access is often limited due to the border with Israel. Making your way to Taba and the Gulf of Aqaba along the border could be good for finding those wintering species like Saker falcon, imperial eagle, houbara bustard and all those sandgrouse which have crept over into Egypt.

Moving directly south of El Arish is the vast wadi of the same name, **Wadi el Arish**, where recently species like desert finch, Finsch's wheatear, desert warbler and Dunn's lark have been found. The area has mainly been covered in winter but with desert finch and Dunn's lark both breeding, it is worth a visit in summer as well. Hawfinch was found here on $12^{th}$ January 1990 and 160 chaffinches in February 1998.

When water is found in the wadi, sandgrouse have been found here including pin tailed, spotted and black bellied sandgrouse. Migration holds many warblers, shrikes, pipits and larks but, with few birdwatchers around, seem to be found on Zaraniq and not here. This, like many other areas in Egypt, deserves better coverage and with improving accommodation in the area why not give this corner a try?

**The main accommodation** is found at El Arish with 1 five-star hotel, 3 three-star hotels and several others of mixed standard. Development to the west of here may include hotels in the future but do not forget the beds at Zaraniq itself. The train is, sadly, no longer running and would have been a great way to view this beautiful area but the buses still run from Cairo and take about 5 hours. There is also a plane which flies two days a week from Cairo, and numerous taxis. Market day is not to be missed on Wednesdays for buying Bedouin goods.

## Western Desert Oases

With 90% of Egypt classed as 'barren desert' these green oases offer the chance of travelling through 100s of kilometres of desert to arrive at one of these often picturesque towns. Over 2,000 sq kilometres of oasis are found at **Bahariyya Oasis**.

The oasis is made up of several villages with the largest being Bawiti with a population of 30,000. With bird-life often seeming very limited so far out in this desert area a ringing scheme in the oases caught an amazing variety of species like wagtails, many warblers including river, savi's, marsh, subalpine, Ruppell's and barred warblers, whitethroat, lesser whitethroat, hoopoe, golden oriole and red backed shrikes. Wetland birds recorded include cattle and little egret, grey, night and squacco heron, lapwing, grey plover snipe, greenshank, little stint and water rail.

Wintering marsh harriers, blackbirds, water pipits and bluethroats have also been seen. Desert, bar tailed desert, hoppoe, Temminck's horned larks, desert wheatear and desert warbler have all been seen along the road to and from Cairo from here.

**Accommodation** and restaurants here are limited. You can travel the 330 kilometres from Cairo by car, taxi or bus. There are hot and cold springs here used in the past by the Romans and many are now not suitable for bathing by tourists. A hard road from Bahariyya to Siwa is found here but there is no public transport yet on the extra 425 kilometre desert journey, so most people travel via Marsa Matruh to reach this oasis.

Only 163 kilometres from Bahariyya is the oasis of **Farafra** which is the smallest of these covered in the Western Desert. Few bird records come from here with eagle owl, marsh harrier, blackbird, white tailed plover, black winged stilt, green sandpiper, cream coloured courser, crowned sandgrouse, Temminck's horned lark, roller, wryneck, isabelline wheatear, water pipit, blackbird, house and sand martin but little time has been spent here by birdwatchers. The hot springs are much better than at Bahariyya!

**Accommodation** is limited here and so too are the restaurants but one new hotel has just been finished, with others planned for the future.

Next is **Dakhla Oasis** which is another 311 kilometres from Farafra and 420 kilometres to the Nile. It is made up of several villages and towns with Mut being the chief town. There are a number of spillover reservoirs from the many irrigation canals in the area and also well pools and salt marshes which are scattered over an area of 155 kilometres long and 10-20 kilometres wide.

Several records came from a reservoir near Mut in the centre of the oasis where up to 300 little grebes have been found as well as black necked and great crested grebe. Also seen here have been cattle and little egrets, grey and night heron, flamingo, shoveler, teal, cormorant and coot along with several waders including up to 500 + black winged stilts, 4 white tailed plover, common snipe, greenshank, curlew and green sandpiper, ruff, curlew and little stint.

A sociable plover was found in the oasis on 22$^{nd}$ March 1993 and other birds recorded in the area include wintering black kite, marsh harrier, long legged buzzard, osprey, kestrel, peregrine, water rail. Migrants include bee-eater, roller, swallow, nightingale and pied flycatcher with many more to find.

**Accommodation** is very limited but a tourist village which will be three or four-star and will include its own restaurants is being built. Car, taxi or bus are the options here for travel with some flights from Kharga Oasis, next-door, on two days of the week.

The **Kharga Oasis** is in a developing stage with the chief town, Kharga, with presently 100,000 inhabitants. Being only 340 kilometres from the Nile and Luxor this oasis has caught its fair share of birds moving through the country with night, squacco and grey herons, lesser kestrel, cream coloured courser, bee-eater, hoopoe, wryneck, swallow, house and sand martin, yellow and white wagtail. Resident eagle owl, hoopoe lark, white crowned black wheatear, palm dove and brown necked raven are all present.

**Accommodation** is again limited but the transport has an added bonus with flights from Cairo on two days of the week. Car, taxi or bus are also available as a means of visiting the area, with trains from Qena.

As with all these oases, coverage is limited and many more species could be found. With the 'New Valley' water scheme about to bring 5% of Lake Nasser's water to this area, great changes will happen in the future and what was once an oasis may well become like another length of the Nile itself. In the south west corner of the desert is the highest point in the Western Desert called Gebel Uweinat, rising to 1,892 metres. It has some interesting birds such as desert sparrow and desert warbler but it is far off the beaten track for most birders.

## Suez

This town is a port of call, not only for the large amount of shipping and pleasure boats using the Suez canal, but also for a wide variety of birds. Some of these species have come from afar and stayed to live here with the most famous one being the Indian house crow which is breeding well here and is now moving to colonise other areas of the Gulf. The present population is thought to be around 4,000 birds, with nests found even in cattle egret colonies.

Birdwatchers have been coming to Suez for a few decades to witness one of the most spectacular movements of birds of prey anywhere in the world, with over

100,000 birds moving in spring and autumn. In spring, recent counts have been carried out mainly by the youth hostel while in the autumn, movement is more likely to be stronger over Port Taufiq.

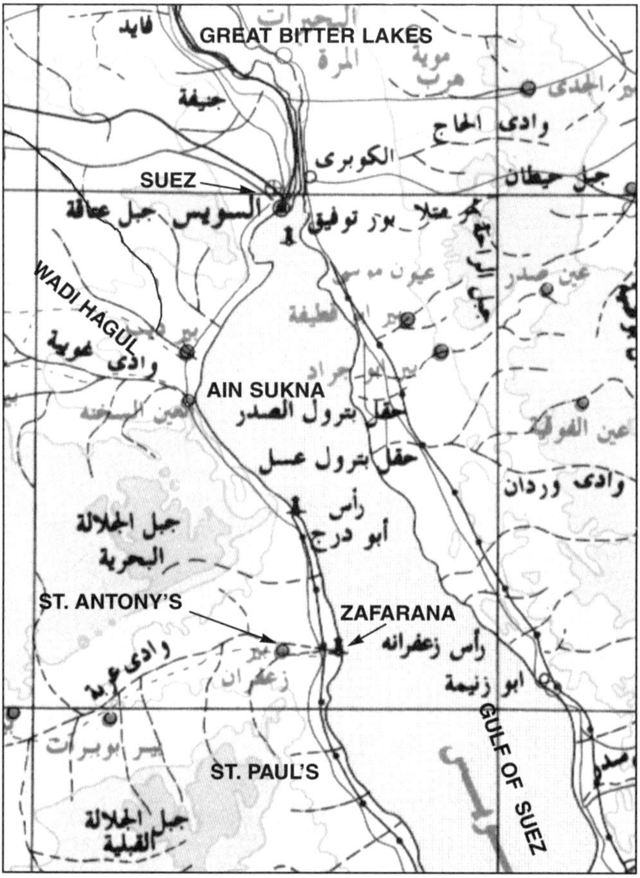


Over 27 different species of birds of prey have been identified with, by far, the commonest bird being steppe buzzards with 50,000 to 80,000 birds counted in some springs. Eagles can total 60,000 of six species followed by much smaller numbers of kites, vultures, harriers, ospreys and falcons of eight species. With so many eyes to the sky, it is amazing that anything else is found, but large numbers of passerines, gulls, terns, waders and even skuas pass by this area of the coast.

The waders are finding it harder each year as Suez Bay is filled in for development. Their important feeding grounds are removed, so fewer will be able to put on weight allowing them to complete their migration. Up to 70 species of birds have been seen using the bay area with 32 species of wader making up nearly half of them. Most of them are on migration and using the area as a stop-over site to feed. Dunlin can be the commonest bird with up to 2,000, with smaller numbers of little stint which have declined quite dramatically over the years.

This is one of the few sites in Egypt where you can find bar tailed godwit on passage and along with it may be black tailed godwit, collared pratincole, greater sandplover and even lesser sandplover in 1990, terex, broad billed, curlew and marsh sandpiper, avocet and oystercatcher. Many herons also pass through here and even western reef heron can be found here. Both black and white stork move through and cranes.

Your first white eyed and sooty gulls may be here as well as many terns including lesser crested and crested, Caspian and white winged black terns. The passerines can be from red throated, tawny or tree pipit, wryneck, penduline tit, nightingale, marsh, great reed, barred, Bonelli's and willow warbler to masked shrike. A brown throated sand martin was found here on 20th April 1990 while counting a large passage of swallows and martins. Access to the bay is limited to the Suez-Port Tawfiq road and the green beach 8 kilometres south of Suez.

**Accommodation** is limited in Suez to 3 three-star hotels and 14 others of a mixed standard, with plenty of restaurants to choose from. The buses from Cairo take only 2 hours to cover the 133 kilometres and both cars and taxis are available from both towns. There are at least six trains a day to and from Cairo and also a boat to Jeddah in Saudi Arabia. This is one of the pilgrim ports for the annual gathering at Mecca.

Using Suez as a base, some of the other sites worth visiting include Wadi Hagul which can be seen by taking the round route from the main Cairo road and coming back along the Gulf of Suez (see also Cairo). **The Great Bitter Lakes** which are now part of the Suez Canal can be found only 15 kilometres to the north of the town, stretching north in two lakes for 40 kilometres.

A small number of waders use these lakes with dunlin, little stint, sanderling, wood sandpiper, lapwing and Kentish plover. Both Kittlitz's plover, marsh sandpiper, and greater sandplover have been found here and even great snipe has been seen. Three species of grebes can use the lakes and several herons are found along the edge, as well as bittern in winter.

Gulls found around the lakes include Armenian, Mediterranean, little, slender billed and yellow legged gulls along with little, Caspian and sandwich terns. Several duck use the area but not in any great numbers with pintail, shoveler, shelduck and wigeon the main species as well as purple gallinule, water rail and little crake.

South of Suez is **Ataqa** where up to 70 house crows can be found, even throughout the day. This site is also used for raptor-watching, both during the spring and autumn months, and south from here is **Ain Sukhna**, where spring raptor-watching can be even better.

The village of Ain Sukhna is not just famous for its raptors as the mountains come right down to the sea, but also for the hot springs which come out of Mount Ataka, the highest mountain around this area of the Red Sea. The raptors are looking for a crossing in spring to get to Sinai and on into Asia or Europe. There are several points from which to watch, even by climbing up the mountain to meet them coming along the ridges. Early in the morning many birds can be found even low down, sometimes sitting out in the desert before reaching the village.

In April 1997 a small area of trees by an encampment held several black kites, steppe buzzards and even steppe eagles, viewed from the road, sitting out in the desert. The heat had not yet risen enough to allow them to carry on with their migration. A small area of water used to be found by the Ain Sukhna Hotel which allowed even birds of prey to come down and drink in front of you. The reed beds are still here but the water is never guaranteed. Large numbers of white storks can be seen here as well as black storks, pelicans and vultures.

The petroleum centre has to be informed of your movements if you are birdwatching in this area, and it is advisable not to have cameras at this point.

Cover around the village of Ain Sukhna has also provided several passerines in the past with red breasted flycatchers, red rumped swallows, olive tree warblers, thrush nightingale, masked and woodchat shrikes. The hotels here will offer even more cover in the future for passage birds. Sea-watching has been good here as well, with species like kittiwake which is a hard bird to find in Egypt but all the locals are here with sooty and white eyed gulls, and passing slender billed and black headed gulls along with several species of tern.

The **accommodation** is new and now allows a stay in this area with 3 hotels all aimed at the sun-lovers, and birdwatching seems a great diversion as the area offers nothing else. If you do not want to stay at Ain Sukhna, Suez is only 55 kilometres north and there are several buses per day as well as many taxis on hire to this site.

Moving south along the road you should see mourning wheatear with smaller numbers of hooded also being reported. The raptor passage can be seen from high in the mountains to low across the sand-dunes, once the mountains have been left behind. At **Zafarana** the dunes here offer feeding for flocks of short toed larks, wagtails, pipits and wheatears. Both crested and Caspian terns were viewed offshore with the beaches sadly covered in oil, while Kentish and ringed plover tried to feed along this polluted shore.

A road goes off to Beni Suef from Zafarana back through the mountains and back to the Nile. This is another chance to get back into the mountains and search for raptors and with the **Monastery of St Anthony** close by, you can drive up and start watching. A record count of 7,280 cranes passed through these mountains on 14th March 1996. Farther down the coast road is another turning to the **Monastery of St Paul.**

The approach was very exciting on 1st April 1997 with several steppe buzzards and black kites flying very low over the wadi. This was followed by a long legged buzzard sitting by the road and then there was a pair of bar tailed desert larks feeding above the wadi. A masked shrike was feeding close to a lesser whitethroat while still the buzzards kept moving through.

Driving through the monastery gates, you could see large numbers of raptors spiralling around the top of a cross on one of the many peaked ridges. A shady place was picked up by the monastery itself and, in just over an hour, several thousand raptors of many species moved through. There was far less variety of eagles than had been seen farther west into the mountains, with steppe being the major one, with even short toed coming second, booted eagle third and lesser spotted fourth. There were no griffon vultures but plenty of Egyptian and there were more black storks than white ones which came in singles only, on that day, with 32 blacks.

A peregrine falcon was unexpected here and so too was the number of passerines around the limited vegetation around the monastery. Both masked and woodchat shrike were joined by orphean, olivaceous, Ruppell's and Bonelli's warblers. Rock martins flew around the monastery while swallow, house martin, yellow and white wagtails passed through. A shop next to the viewing site was convenient in such heat with continuous visits to keep up the liquid.

**Accommodation** here even has separate rooms for men and women, and food is also provided. Like so many of the sites in Egypt, little coverage is given to these monasteries and there must be many more records to come in the future.

Bee-eaters - European, Little green, Blue-cheeked.

Marsh Harrier

Senegal Coucal

Painted Snipe

Spur winged Plover

Wadi Hagul

Port Said

Great white Egret

Little Bittern

Night Heron

Great Snipe

Lake Burullus

Kentish Plover

Yellow Wagtail (blue headed)

Golden Oriole

Water Pipit

Stonechat

Whiskered Tern

Fort Qait

Montaza

Montaza Harbour

El Alamein Commonwealth Cemetery

Cultivation

Hoopoe

Little Crake

Green Sandpiper

Mediterranean garden

Marsh Sandpiper

Black eared Wheatear

Lesser grey shrike

Whinchat

Great reed Warbler

Black Redstart

Siwa Oasis

## Hurghada

The town is not named after its abundant coral, fish and birds but after a salt shrub, *nitaria retusa*, called ghardaka in Arabic, which grows in the area to help keep the small sand-dunes together. One of the great attractions of this town is that you can combine the great hobby of diving with finding birds; not, of course, at the same time but on the boat journey to the dive sites, first thing in the morning or, on your return. The benefits can be very rewarding as, firstly it fills in the spare time, and secondly it may encourage travel along the coastline rather than being confined to the town.

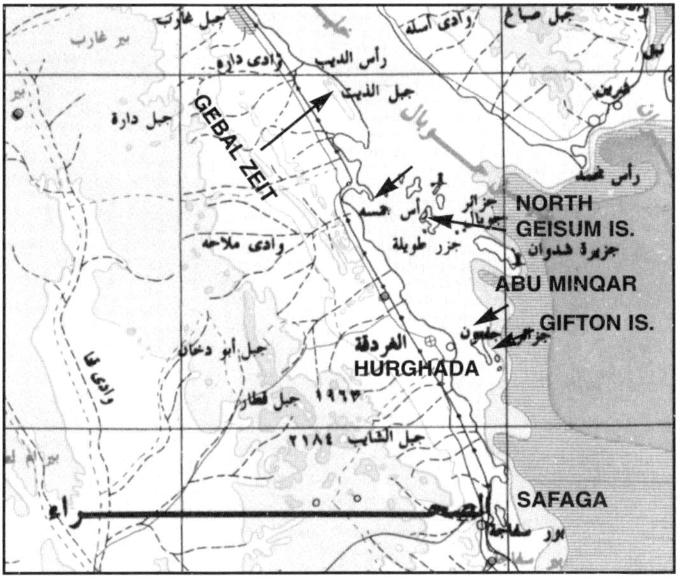


## One interesting two-week break went as follows:-

The trip was from 7th to 20th April 1995 and the dates were picked to coincide with the school holidays in England. This is also migration time for the birds so some of the records were due to this passage.

The couple in question flew direct to Hurghada, landing at 3pm and started finding birds even at the airport with northern and isabelline wheatears, whinchat, tree and red throated pipits and lesser whitethroat. They drove to the hotel (Friendship Hotel), settled in and then started birdwatching around the hotel and adjoining grounds. Here they found several yellow and white wagtails, palm dove, woodchat shrike, swallow, more pipits, egret species flying over, chiffchaff, tawny pipit, marsh warbler and a pratincole. All were seen within a few minutes walk of the hotel.

Next day before breakfast, several birds including nightingale, sedge, willow and reed warbler, redstart, blackcap, more tree pipits and northern wheatear were accounted for. Then it was onto the harbour to start the real holiday, with white eyed, sooty and slender bill gulls seen along with cattle egret, Caspian and crested terns. This being the first dive, many fish were not identified but 30 species were, along with several corals and sea slugs.

New birds on the next day included ortolan bunting, rufous bushchat, olivaceous warbler and greenshank along with many of the others seen the previous day. From the boat many of the same sea-birds were seen but some distant small terns were not identified. Several new fish were added to the list that day, and an evening walk along the shore added both ringed and Kentish plover.

Where the vegetation is mature and diverse, many more birds can be found, so a trip was made to the Sheraton Hotel's grounds and there were 5 Levant sparrowhawks, Spanish sparrows, quail, 3 wrynecks, pied kingfisher, subalpine warbler and many more of the previous day's birds. More Levant sparrowhawks were seen from the boat, along with the usual sea-birds.

The star birds of the next day, onshore, included a male marsh harrier moving along the coast, along with several species of martin and swallows including sand and house martin and red rumped swallow. Just before the boat departed, a western reef heron was spotted along the shore and a Montagu's harrier was seen flying over the sea. The first white cheeked terns were identified along with little terns out to sea.

Next day, it was back to the Sheraton Hotel where more Levant sparrowhawks were seen along with another ortolan bunting. Black eared wheatear was a new bird as was a flock of 8 blue cheeked bee-eaters. There was also a superb collared flycatcher and several wood warblers. On the way to the harbour there was grey heron along the beach with several species of gulls and terns.

No new birds were found the following day but bottle nose dolphins were seen from the boat and the next day provided the first and only view of brown booby out to sea. A few days were taken away from the town via a local bus ride to Luxor where the Nile provides several new species of bird, with the best record being 7 lesser kestrels flying around a mosque. Then it was back to diving and lesser black backed gulls. Ramada Island provided 3 ospreys and a striated heron.

The next day was an early morning taxi ride to **Bir Beida**, west of Quiser where the famous crowned sandgrouse can be seen flying to drink from the wadi. There was an amazing 300 on that day. Also there was a passage of raptors including sparrowhawk and Levant sparrowhawk, male Montagu's and 3 marsh harriers, steppe buzzards and 6 steppe eagles, 10 black kites, osprey and 2 lanner falcons.

A white stork was feeding in the wadi along with several species of yellow wagtail including black headed, Egyptian, blue headed and others. A savi's warbler was singing in the reed bed and several green sandpipers were feeding along the water.

The last day brought a few goodies with hoopoe, turtle doves, spotted flycatcher, 7 collared pratincoles, European bee-eaters, ruff and was it a semi collared flycatcher? With the extra trips and local birds combined, just over 100 birds were seen on this 'diving' holiday and not forgetting some of the world's most stunning fish and corals, and plenty of sunshine.

With the increase in hotels come the tourists and with the added cover comes the bird records. The numbers of warblers, pipits, shrikes, waders and even birds of prey are increasing all the time and anyone staying for a short time can come up with 100 species of birds during the migration time. All this is accidental man-management and on this same theme comes the **Hurghada rubbish tip**.

Found to the west of the town the rubbish tip is an expanding piece of land which has been taken over by the birds, especially during migration. Both water and food are found here so the birds stop over to build up their body fat to help them on their way. Up to 15 species of wader were using the area on 2nd April 1997 with white tailed plover, black winged and collared pratincole, marsh, wood, green and common sandpipers, ringed, little ringed and Kentish plovers, little and Temminck's stint, black winged stilt, greenshank and ruff.

**W**hiskered tern was hunting over the water while grey heron, shoveler and ferruginous duck fed by the side. Quail was found along with tree and red throated pipit, northern and black eared wheatear. Lesser short toed larks were joined by crested and short toed larks while several flava yellow wagtails, mixed with whites, made a great few hours' birding. The world's largest concentration of white eyed gulls featured on the rubbish tip, with 6,500 here on 7$^{th}$ May 1996.

**S**haring the beaches with the tourists can be several good birds with records of both greater and lesser sandplovers, crab plover, broad billed sandpiper and Caspian plover. Locals include western reef heron, white eyed and sooty gull and osprey with lanner and Barbary falcons in the area. Small flocks of sooty falcons have been seen over the town, catching insects at dusk but it is out to the islands when things get really hot.

**T**he main tour which nearly everybody does from here includes a 1-2 hour trip out past Abu Minghar Island with its breeding reef and striated herons (Goliath heron has been seen here in the past). White cheeked, Caspian and lesser crested terns are often seen around the island and the anchoring site is between Gifton and Little Gifton islands where you can go snorkelling. You are often accompanied by white eyed and sooty gulls and, in season, there should also be white cheeked terns close by.

**A** few hours are allowed on Gifton Island where white eyed gulls and white cheeked terns nest behind the beach where the tourists swim. There are also ground-nesting ospreys here as well as Caspian terns on small coral islands in the bay. The boats do not give time to cover the full island but sooty falcons have been seen at the northern end of the island and even cream coloured coursers, Kentish, ringed and little ringed plovers, turnstone, isabelline and black eared wheatears, wryneck and brown necked raven have all been seen. Two long billed pipits on 26$^{th}$ September 1996 were the first records for this area.

**F**rom the boat ride you can often pick up several tern species and even raptors travelling across the open water as they head for Sinai. Brown boobies are easier to spot in winter than in summer when many have moved to their breeding grounds farther north, and red billed tropicbirds seem to have stopped breeding in the area with only odd records each year of birds moving amongst the islands. A real rarity was South Polar Skua on 10$^{th}$ June 1991 but other skuas have also been seen, including pomerine skua.

**B**oth Um Magawish Islands have a good number of birds around them with bridled terns more likely to be seen into May, while great black headed gulls are only found in winter with most birds gone by April. Spoonbills have bred in the past and can be seen occasionally in the mangroves off the islands. This group of islands is much easier to see from Hurghada, while moving north the islands here are too far to travel for most of the boats based here.

**T**hese northern islands still have many breeding birds, with **North Geisum** holding the largest breeding population of brown boobies in the area. The breeding birds are very much the same as the ones found off Hurghada but sadly, there are no more records of red billed tropicbird breeding here. Access is limited at the moment but with the increased development of hotels 20 kilometres to the north of Hurghada, boats may well visit these islands from here.

**B**ack onshore the hotels worth checking for migrants are the Sheraton, Jasmine Village, Magawish, Arabia Tourist Village and the Movenpick. Many others will improve with time as their plantings mature. Other birds not mentioned from here include olive tree warbler, all three species of pied flycatcher, red breasted flycatcher, Siberian rubythroat, citrine wagtail, bimaculated lark, and river warbler. Large numbers of cranes, storks and herons can move across the country, heading for Qena and the Nile and returning with some travelling along the coast via Hurghada.

**Accommodation** is widespread with over 7 five-star hotels, 12 four-star, 12 three-star and over 35 other standard grade hotels. There are direct flights from many countries which saves on the 510 kilometre journey from Cairo. The buses can take 7 hours from Cairo and there are 10 or more a day. You can also get to Luxor from here which takes 4 hours. Taxis are everywhere and car-hire can be organised through many hotels. A ferry can also be taken from here to Sharm el Sheikh if you want to divide the holiday. The ferry takes 4-6 hours and leaves in winter on Tuesday, Thursday and Sunday and every day in summer, except Friday.

As many of the hotels are named under Hurghada, you have to be aware that you could be based at a distance of anything up to 20 kilometres from the centre of town. Most large hotels do have a free mini bus service if this is the case, but it can be very misleading to arrive and find out that the restaurants are not on the doorstep!

As a centre, Hurghada takes some beating with access to other parts of the country and along the Red Sea itself. Tourist destinations can also offer the chance of bird sightings with a ride to Mons Porphyritis being a chance to see desert birds such as bar tailed desert lark, cream coloured courser and sand partridge while spring and autumn raptor migration may also be seen here.

A trip to **St Paul's** or to **St Anthony's Monastery**, likewise, may add on the birds but a visit to **Gebel el Zeit** (89 kilometres from Hurghada) needs pre-planning as there is both a military presence and an oil terminal here. Passage is best seen in the spring as birds try to reach the Sinai over this area of high ground. The winds play a major part here with some days birds sticking to the mountains and others when birds come down over this rise.

Some counts of over a thousand birds have come from here and include steppe eagle and steppe buzzard, honey buzzard and Levant sparrowhawk. Other birds have included pallid harrier, red kite, imperial eagle and lanner falcon. Until this area is more open to birdwatchers there seems little need to cover it and time would be better spent back in the mountains where the bulk of these birds came from. **Ras Gamsea** (59 kilometres from Hurghada) is another raptor passage site with easier access where also several waders have been found, including greater sandplover, collared pratincole, grey plover and curlew sandpiper.

Under development at the present time (1998) is a large holiday complex south of Hurghada which plans to create its own fresh water by building a de-salination plant. Here, even a golf course is planned in the desert so future birdwatching may add yet more birds to this area.

Down the coast is **Safaga** which is a port mainly for transporting phosphate from the area, and a pilgrimage port for boats to Mecca. There are a number of hotels here which again can catch the migrants as they move along the coast. Species like redstart, pied flycatcher, wood warbler, lesser grey and masked shrike have been seen here but most people are passing through on their way west towards the Nile. House crows have also made it this far.

This Safaga-Qena road is also carrying the main water pipe from the Nile to keep all these hotel swimming pools full. This pipe has several leaks and many migrants and desert birds have come down to drink along it. This road is a good place to see spotted sandgrouse along with hoopoe, desert and bar tailed desert lark, sand partridge and even cream coloured courser. Raptor passage can be observed along the high ground.

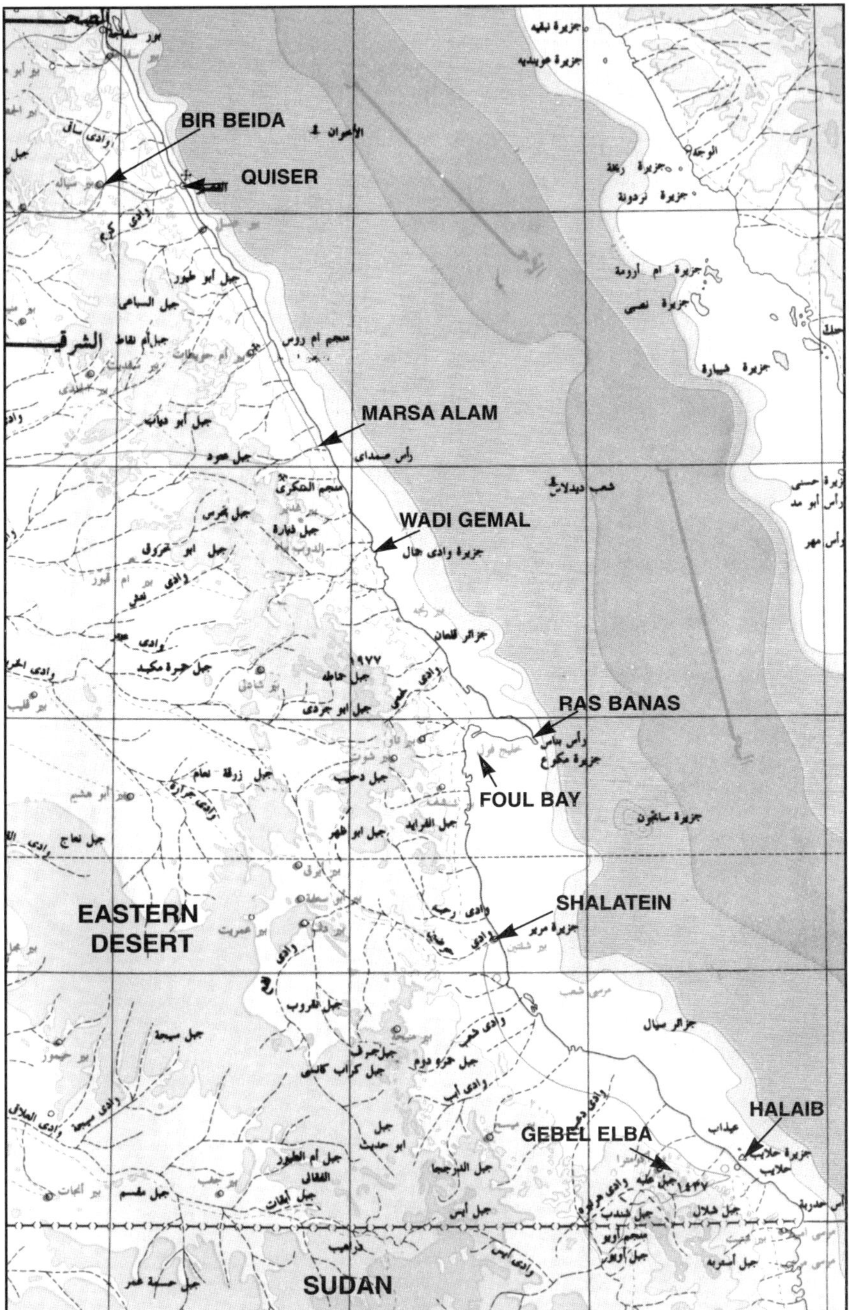
BIR BEIDA
QUISER
MARSA ALAM
WADI GEMAL
RAS BANAS
FOUL BAY
SHALATEIN
EASTERN DESERT
HALAIB
GEBEL ELBA
SUDAN

Just south of Safaga is the first mainland mangrove swamp where western reef heron is common, along with spoonbill. Goliath heron has been found here along with visiting greenshank, common sandpiper and great white egret. Wryneck has been found here with many flava yellow wagtails and at sea, Caspian tern and slender billed gull have been seen.

The journey between Safaga and Quiser is limited in cover other than these mangrove swamps and one wadi which has a large covering of tamarisk. In April of 1997 ospreys, spoonbills, western reef herons, Caspian terns and steppe buzzards were all seen along this road. **Quiser** itself is very small with only 3 hotels, one of which is a Movenpick - it will not be long before the birds move in! A record of black wheatear came from the town but there is always some concern that the species could be confused with white crowned black wheatear. You can sea-watch from the town but the main birding spot lies west along the Quiser-Qift road at **Bir Beida**.

Here, over the years, a remarkable variety of birds has been found and fresh water is the key. A spring runs out of Wadi Abu Ziran with reed beds and open water surrounded by high mountains. Rarities include pale rock sparrow, isabelline shrike and close by, crested honey buzzard and lapped faced vulture. The regular event is the gathering of crowned sandgrouse with up to 500 birds having been counted. The setting is wonderful as their calls echo around the wadi.

Trumpeter finches are another regular here, coming down to drink at the water, and migrants vary from green sandpiper, little egret, white stork, marsh harrier, short toed, steppe, Bonelli's and lesser spotted eagle, steppe and long legged buzzard to great reed, savi's, marsh and sedge warbler in the reeds as well as Ruppell's, Bonelli's, willow and olivaceous warblers. Spanish sparrows, bluethroat, Sardinian warbler, chiffchaff and lesser whitethroats can be found in the winter.

The **Quiser-Qift** road itself is also good for locals such as hooded and white crowned black wheatear, sand partridge, spotted sandgrouse and brown necked raven. Migrants have included bee-eater, masked shrike, golden oriole and wryneck.

Travelling south again from Quiser is another mangrove bay with a new hotel named after it, perched overlooking it. Again Goliath heron has been seen here along with the other normal species for this area. A considerable amount of planting has taken place around the hotel and again, migrants will be drawn into its grounds.

At Marsa Alam a new airport is planned to bring tourists directly to this part of the coast. At the present time there is no access south from Marsa Alam so the development of tourism may be the only way in the future that will allow better access for birdwatchers to this area, with Hurghada now 275 kilometres away.

The **Marsa Alam-Edfu** road has allowed finds such as hooded, mourning and white crowned black wheatear, cream coloured courser, sand partridge and desert lark, with raptors during passage. Farther south, the Wadi Gemal area has mangroves which hold breeding, western reef and striated herons. The whole area has been designated for development so, although access will be allowed, it is not yet clear what will happen to the wildlife!

Farther south at **Shalatein**, you have the capital in Egypt for crab plovers with over 200 being seen here over the years, with 38 on 30th March 1997. There were also 15 lesser sandplovers here on 1st April 1997 along with 2 lapped faced vultures on 30th March 1997. Coastal islands here have breeding red billed tropic-bird and brown booby, with many turtles using the beaches to breed. You are now within striking distance of the 'sacred' Gebel Elba region which holds a number of breeding birds found nowhere else in the Western Palearctic.

The habitat here is unique to Egypt, with mountains which actually have trees and shrubs growing on them. The two special birds found nowhere else in the Western Palearctic are shining sunbird and rosy patched shrike, with many more which breed here but have been seen elsewhere. These can include ostrich, Verreaux's eagle, bateleur, Arabian warbler, pink headed dove, Nubian nightjar, long billed pipit and black crowned finch lark along the coast.

Fulvous babbler is now found only here after losing its breeding site along the Nile before the area was flooded, while some species have an isolated breeding population down here. Most of these, including fan tailed raven, Hume's tawny owl, Lichtenstein's sandgrouse and blackstart have other populations in Sinai, and whether this isolation is due to the destruction of habitat between here and Sinai is another story!

Some species have been found here but it is not known if they breed only occasionally and these include Sudan golden sparrow and African silverbill. Birds like common bulbul are isolated here but found in other parts of Egypt. Very few birdwatchers have had the chance to explore this area and new breeding birds could be found in the future. Very little work has been carried out recently in winter to look at what is here and this makes the area one of the few uncharted areas of the world!

Future access is unsure and more than likely, it will be through specialised bird groups/tours if access is allowed at all. Certainly the coastal areas need wider coverage and this may produce records of brown noddy, Saunder's little tern and many shearwaters and petrels. Certainly an area well worth waiting for!

## Upper Nile

## Luxor

Although this area has already been mentioned under the cruising, there are several parts which can be covered by non-cruisers. Most birders will look at ancient monuments even though they may not be wholly interested in them and to tell them that they can see birds there seems to be an even bigger incentive. Starting at Luxor, the east bank offers the Karnak Temple and the Luxor Temple with a wide variety of birds, but if you prefer the shade, the Winter Palace gardens should not be missed at any time of the year.

On the West Bank the Valley of the Kings is always very hot and is well worth visiting early in the morning, for local species like trumpeter finch with up to 200, hooded, mourning, desert and black eared wheatears, and desert lark. Raptors and storks usually come later but you need both shade and water if you plan to stay and watch for passage. Both Egyptian and griffon vultures have been seen as well as steppe, lesser spotted, short toed and booted eagles, black and white storks, cranes and honey buzzards.

The west bank of the Nile has easier access into the agricultural land and if you take a felucca to cross the river also try a ride up to the islands to the south of the town where a large number of herons and egrets can be found. This is a good way to obtain pictures of the birds. The first island can hold night, squacco, purple, grey and striated herons as well as little egret, purple gallinule, moorhen, pied kingfisher, clamorous reed warbler and even hunting marsh harrier.

Passing Crocodile Island you start finding the large duck concentrations with shoveler, wigeon, teal, pintail and garganey with recent records of ferruginous duck and marbled teal. Egyptian geese are here in winter with both white winged, whiskered and gull billed terns. With a trip to the north you have to reach Khozam if you want to have any real chance of seeing the African skimmers but I have even seen them twice, flying through the middle of Luxor!

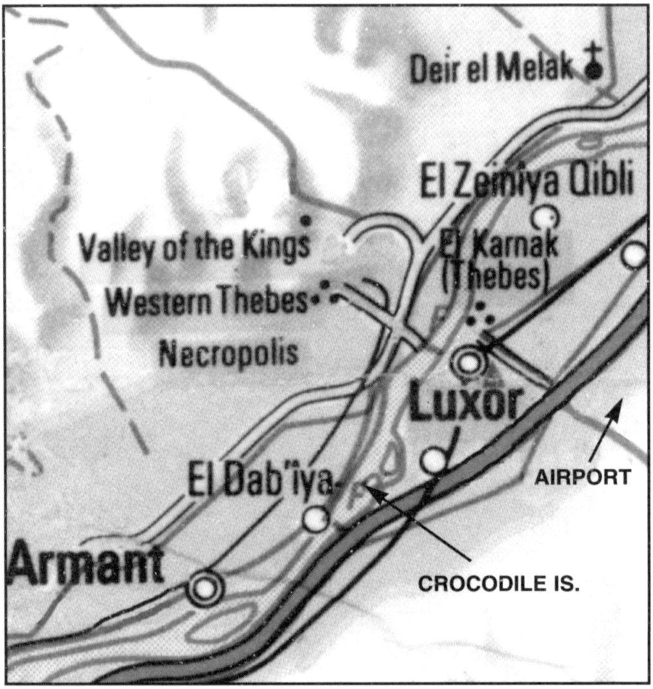


The agricultural land here has most of the local birds with spur winged plover, little green bee-eater and fan tailed warbler. Migrants include several species of warbler, nightingale, golden oriole, wryneck and both woodchat and masked shrike. The area which has received most coverage over recent years is the grounds of the **Crocodile Hotel** which is 4 kilometres south of Luxor - over 150 species of birds have been identified.

The combination of good hotel policy and an ideal situation for the birdwatchers offers a wonderful experience. Instead of treating all those nasty bugs with pesticides why not let the bird-life sort it out for you? The resident birds are amazing, with my room at F8 being special in receiving a displaying hoopoe at first light. Non-birders would have been furious to be woken up so early but I sat in a chair by the window and had the male bird inches from me as it tried to ward off that other male it could see in the glass!

The hotel **accommodation** is set out in circular chalets with exotic grounds of many mature trees, shrubs and flowers. Nile Valley sunbirds are everywhere, both palm and turtle dove nest side-by-side, common bulbul, graceful and olivaceous warblers, little green bee-eaters and the gem for most birdwatchers is the breeding rufous bushchat. They are so tame here when normally they are darting for cover.

The vegetable garden is organic and then there is the river with its avenue of eucalyptus. Pied kingfishers dart along the water and several duck including tufted duck, gulls, terns and herons are seen. At the end of the avenue is a reed bed with an adjoining marsh. Here you find your tame purple gallinule, squacco herons and both striated heron and little bittern, not to mention the clamorous reed warblers.

Masked shrikes compete for insects along with the bee-eaters while the hotel grounds drift off into farmland. Here there is a mixture of grass and crops with the river dictating the growing seasons. The water is high in the summer months so most crops have to be out by April. The winter low leaves ideal conditions for wading birds and that is why this site has become such a known site for, especially one bird, the white tailed plover.

In winter, up to ten birds can be found here feeding along with the non-breeding spur winged plover. The best site is at the southern end of the island where large areas of mud can be found along with marsh, wood, green and common sandpipers, redshank, greenshank, dunlin, little and Temminck's stints, ruff, black tailed godwit ringed, little ringed and Kentish plover, snipe and black winged stilt. Spoonbills use the area for feeding along with glossy ibis, many more duck and Egyptian geese. Reed, sedge and moustached warblers use the many reed beds and red avadavat have recently colonised the area.

Both hen and marsh harrier use the area as well as black shouldered kites. Walking back along the channel towards the bridge there are some sandy areas where the local Senegal thick-knees are found. Other good birds from the island include little crake, painted snipe, namaqua dove and marbled teal. Residents can borrow both binoculars and telescopes from the hotel but sadly, few people take up the offer. There is also a check-list of the flora and fauna found on the island in both English and German. Not many hotels go to that extent, even in Europe!

To the east of Luxor the airport is well worth a visit with large open areas where both Montagu's and pallid harrier have been seen along with saker and several sooty falcons. The farmland can have too much sugar cane but in open areas flocks of storks can come down to feed along with many pipits, larks and wagtails.

If Luxor is your base it is very easy to move around, especially on visits to the Red Sea or to Aswan.

**Accommodation** is widespread in Luxor with 5 five-star hotels, 5 four-star, 13 three-star and 19 other standard grade. Luxor is also the major starting point for all those floating hotels with a staggering 108 of five-star rating, 48 four-star rating, 27 of three-star with 32 other standard type. Planes fly direct to Luxor from many countries and you can also fly internally to and from here.

The buses from Cairo go only via Hurghada and take between 10 and 12 hours while the buses travel the 209 kilometres to Aswan in 3-4 hours. The trains do run along the Nile and can take up to 10 hours to Cairo and the shorter journey to Aswan is very much more pleasant. There are plenty of taxis in Luxor and with all those flat roads by the river some birders do it by bike!

The agricultural farmland between Luxor and Aswan has many of the local birds but the main stop-over sites are again the ancient temples with sites at Esna and Edfu where the river can produce a wide variety of ducks and grebes, with a record count of 730 ferruginous duck in 1996.

Moving on to **Kom Ombo** where you find the most records of African skimmer in the Nile itself, with many from cruise ships but also from the land. The main islands or sand bars are on the western bank but are directly opposite the temple at Kom Ombo. Here, up to 43 birds have been seen and closer to the east bank several waders have been found including Kittlitz's plover, greater sandplover and white tailed plover.

Temminck's stint is a regular winter visitor here with records of several warblers, Egyptian nightjar, ruddy shelduck and even namaqua dove. Both bittern and little bittern have been seen in the reeds here and the numbers of pallid harriers have been increasing as you approach Aswan. Great spotted cuckoo have bred using hooded crow nests between here and Aswan.

## Aswan

A town which has become the start of the River Nile in Egypt with the building of the new dams, making Lake Nasser back up into Sudan. It is a very colourful place with rocky islands compared with the normal sandy ones downriver, looking across to a desert dune system. Most tourists will not see anything like these anywhere else in Egypt, and this gives the town yet more charm. It is a great place for observing passing migrants with flocks of waders, duck, storks, pelicans, cranes and raptors, not to mention the ones which drop down to feed right in front of you.

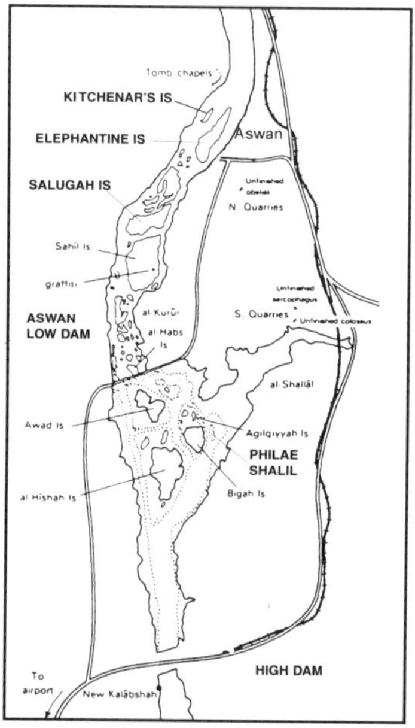


With so much boat traffic in the area it is surprising that so many birds such as numerous flocks of garganey, shoveler, wigeon and teal actually put down in the area. Due to this disturbance, Egyptian geese do not usually stop but pass right on to Lake Nasser but many herons feed in the area along with egrets, and many now remain to breed with both cattle and little egret breeding on Kitchener's Island. Even flamingo, Goliath heron and yellow billed stork have been seen here.

One of the main viewpoints for watching migration is from the top of the New or Old Cataract Hotel. You ask politely if you can take some pictures from the top of the hotel. The numbers of vultures has been very good from here with griffon, Egyptian, black and lapped faced vultures. Falcons have included Barbary, lanner, peregrine, sooty and kestrel with several eagles including steppe, lesser spotted and short toed. In winter, crag martins can be seen flying around and dawn casts a great the sand-dunes to the west.

There are several gardens which are worth checking. The Fryial next to the Old Cataract Hotel being one of the best, followed by a public one next to the Rameses Hotel but after that it is on to the islands with the gardens of the Oberoi Hotel well worth checking with records of Nile Valley sunbird, golden oriole, nightingale, Ruppell's warbler, redstart, collared flycatcher and roller. Kitchener's Island has an amazing arboretum which attracts birds from all over. The breeding egrets also enjoy the trees while on the west side of the island black kites gather on the sand-dunes, with up to 250 counted.

Saluga and Ghazal Island Nature Reserves, upriver towards the first cataract, are some of the few islands left in the Nile which are uninhabited and covered by their natural vegetation. Here birds like squacco and striated herons are breeding along with night heron, and a juvenile little bittern has also been found here. In winter, the blue streak of kingfisher is a common sight along with migrant grey and purple heron. A small crake of either Baillon's or little was found here in July suggesting that it was a local breeding bird. Feluccas commonly come this far up the river and this can be a wonderful experience.

The farmland around Aswan is often a staging post for migrating birds with flocks of cranes, white storks and glossy ibises stopping off here. In April 1997 the first breeding record of namaqua dove was found here and there are good numbers of shrikes with blue cheeked and bee-eater on passage. Great spotted cuckoo is well worth looking out for with some birds breeding locally.

The scrubland opposite Philae is worth a look with rufous bushchat, Nile Valley sunbird, Egyptian nightjar and vulture, with long eared owl in 1997 and masked shrike with the water areas offering greenshank, whiskered and gull billed tern and past records of white tailed plover. The pelicans here are well worth checking with occasional records of pink backed pelican. Many diving duck use this lake between the two dams and several records of ferruginous duck come from here. Ospreys are a regular here with at least a single bird perched somewhere around the lake. The high dam can often be too hot for much to happen but vultures are a good bet.

**Accommodation** here is with 5 five-star hotels, 3 four-star hotels, 2 three-star hotels and 13 of standard grade accommodation but do not forget those floating hotels with 108 of five-star rating, 48 four-star rating, 27 of three-star with 32 other standard type. There are also a small number of floating hotels on Lake Nasser.

Flights come from Cairo but you can also get down to Abu Simbel, with one company talking about flights to Khartoum in the Sudan. That would result in a new type of birding holiday! But this has not happened yet. There are ferries still running to Wadi Halfa, Sudan from the high dam and taking between 14 and 24 hours with no stops, not even at Abu Simbel.

The train from here to Cairo can take from 16 to 20 hours and is the same price as from Luxor. There are many buses heading both north or south, and taxis are everywhere. If you are heading for Abu Simbel make sure you have plenty of water because this journey is real desert with mirages everywhere, making Lake Nasser look as if it is often flooding the road.

The journey to Abu Simbel is one of those when you either see a lot, or sometimes, nothing at all. There have been records of a pair of cream coloured courser just past the checkpoint where there was also desert wheatear. The first wireless station had a lanner floating over it while an area of cultivation, made by the army just 25 kilometres out of Aswan, held isabelline shrike in 1995. The first lapped faced vultures have been seen close to here.

Around 50 kilometres along the road is an area of desert scrub which has held several orphean warblers, lesser whitethroats and whinchat. This stop also gave away the presence of crowned sandgrouse with 6 seen close to here. Both desert larks have been along the road, as well as hoopoe lark. Migration can show even swallows, moving across the road which seems strange when the lake is not too far away.

The halfway resthouse does have water and several migrants are likely here with flava yellow wagtails, desert and black eared wheatears and larks, and pipits. The next stop for the tourists is often the camel feeding area which also offers weed seed for the local sandgrouse. Over 200 spotted sandgrouse have been seen here along with tawny pipit, short toed lark and even spotted flycatcher. Both lanner and kestrel have been caught by the locals to use as a photo opportunity for money.

## Abu Simbel

By this time you want only to get to your hotel for a shower and you miss several roads leading to new agriculture down by the lake and these will have to be done later - the 280 kilometre journey can take anything from 4 hours to all day, depending on the bird-life. At Abu Simbel your first thoughts are on Africans so, after your shower, it is time to start work with numerous bays offering the chance of anything, with over 190 species having been seen down here.

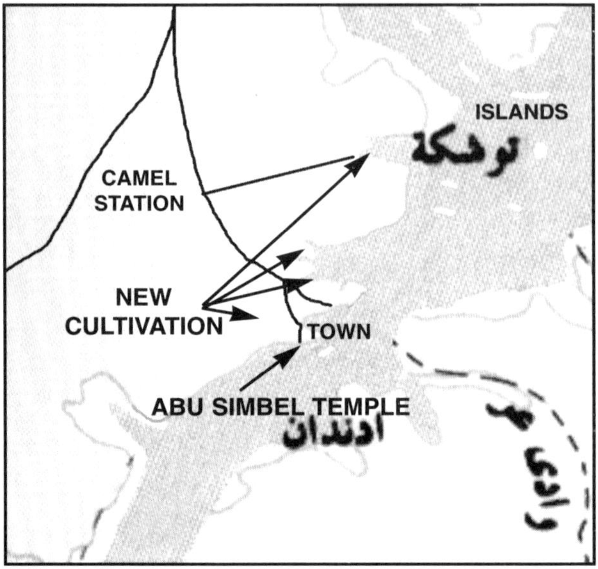


The summer months seem to be the best for quality but the heat and the lack of migrants may put you off. April seems to be the month to get the best of both worlds but anything from August through to March can still be rewarding. The vegetation around the hotels and village hold a number of good birds with several records of both African collared dove and namaqua dove. Large numbers of warblers pass through with surprising numbers of marsh, great reed and orphean warblers.

Winter holds many Sardinian warblers with chiffchaff and lesser whitethroat, with smaller numbers of olivaceous, Bonelli's, sedge and willow warbler. Several wheatears have included pied and Cyprus pied, isabelline, black eared, northern and desert wheatears. Isabelline shrikes have been found with recent records of both striated heron and bittern. There are many grey herons with a scattering of purple but much fewer squacco and night herons, with the odd Goliath heron turning up. Kingfisher is all on its own this far south but numerous birds have been increasing with white pelicans, spoonbills and even flamingos.

A visit to the temples here can produce several warblers and even African pied wagtail and Abyssinian roller. Black kites, brown necked ravens and rock martins are the common birds above the temples but lanner, sooty and saker falcons, steppe, Bonelli's and imperial eagle have all been seen and, on an evening, both Egyptian nightjar and eagle owl feed in the area.

To most birders it is the boat trip which is the highlight here with a visit to several of the islands out in the lake. African skimmer, Kittlitz's plover and African pied wagtail are recent breeding birds, while colonies of cattle and little egret plus squacco heron were found in 1997. Yellow billed storks, pink backed pelicans and long tailed cormorants are all possible migrants.

The numerous rubbish tips can have black kites, cattle egrets, white storks, Egyptian, lapped faced and griffon vultures which are the rarer ones.

The agricultural land here is a magnet for migrants with many red throated and tawny pipits, flava yellow and citrine wagtails, bluethroats, Spanish sparrows and wheatears.

Recent records have included both yellow billed stork and pink backed pelican in the numerous bays here, and an African pied wagtail was holding territory on the shore here, using an old oil drum as its nest site. Kittlitz's plovers breed along the shore and numerous other waders can be found here. Lesser flamingo was found here in November 1992 and lesser moorhen was here in May 1997.

On the lake itself, large numbers of ducks have been found especially tufted, shoveler and wigeon with black necked grebe leaving the northern lakes to concentrate here. Up to 22 great black headed gulls have been counted as well as Armenian, yellow legged, great black backed gulls and pomerine skua. All these records are happening just a few kilometres from the Sudanese border so future records, such as palm swift, Egyptian plover, sacred ibis, Ruppell's vulture, gaber goshawk and fulvous babbler, may well add species which are already over the border.

Increasing agricultural schemes and development are other reasons to attract birds and the high water levels of 1996/97 produced larger concentrations of, especially waders, 20 kilometres north of Abu Simbel. More Kittlitz's plovers were found along with spoonbills, greater flamingos, glossy ibises, white winged black terns, black winged stilts and avocets.

The sheer size of Lake Nasser says that lots go missing and it is predicted that the lake will become a major wintering area for many species in the future as the silts build up and leave better edging effects. The present-day cruises on the lake rarely touch the bird spots with Wadi Allaqi offering several species, the prize being ostrich, eagle owl, crowned sandgrouse and, compared with the Nile cruises, the size of the lake often leaves the shoreline too far away.

**Accommodation** here is sadly limited to only 2 hotels, with 1 four-star and 1 three-star rating hotel. Future development at Toshka may create new accommodation but this factor can limit the time of duration in this area. The planes are aimed at temple seekers and several come each day with tourists but are soon gone again. The buses are quite adequate but taxis here are very limited but some locals will take you out for the day. If in doubt, it may be easier to bring one down from Aswan as the hotels will/can offer the driver free lodging.

## South Sinai

## Sharm el Sheikh

This area offers the tourist and birdwatcher yet another destination in Egypt which can be covered at a leisurely pace. The area seems cut off from everywhere else and can give the feeling of being in another country. The main centre is usually the Sharm el Sheikh area which has the widest choice of accommodation but with further development along the Gulf of Aqaba, even Dahab, Nuweiba and Taba can act as stop-overs.

The history of birding in South Sinai is often better documented than most other parts of Egypt, with increased visits from the 1970s due to the growing number coming over the border from Israel. The two main sites which were visited were St Catherine's Monastery and the sewage works at Sharm el Sheikh. The coastal area is world-known for its coral reefs and it is this that has prompted the massive development of hotels along the coast.

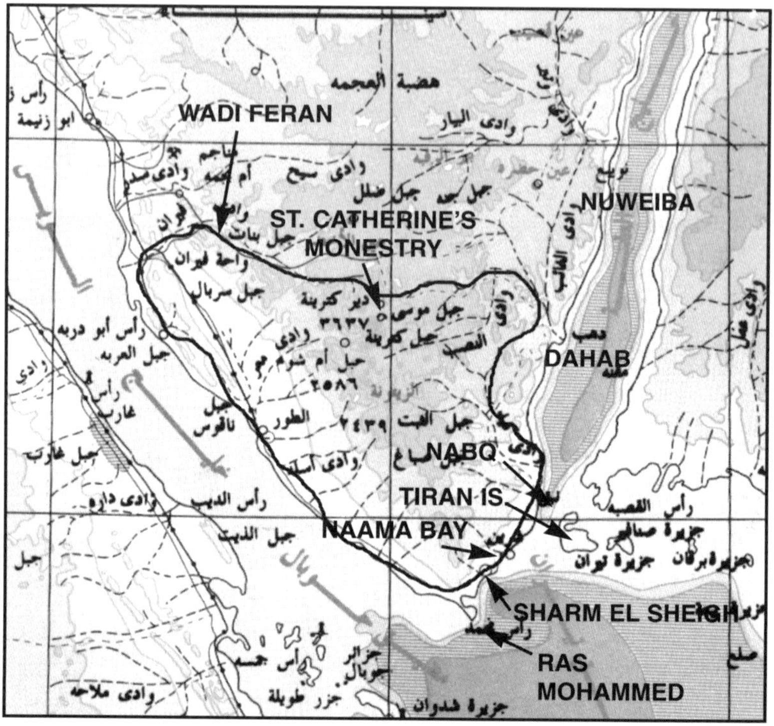


Using Sharm as your centre the first stop is nearly always the area around the **sewage farm** which can be seen well off the main road and the turn-off is next to a service sation. Starting at dawn, crowned sandgrouse (up to 524 counted) and small numbers of spotted sandgrouse come in for a drink to start their day. At dusk it is the turn of the Lichtenstein's sandgrouse with up to 120 birds coming to drink. The water and the maturing trees can bring down many migrants, varying from thousands of storks to a single cinereous bunting.

Oriental skylark has been found along with up to 6 Richard's pipits in 1996 and other records include penduline tit, buff bellied pipit, Menetries, Upcher's and moustached warbler, Siberian stonechat, little and yellow breasted bunting, desert finch, and citrine wagtail. White breasted kingfishers are becoming autumn specials but there are still few records of spotted crake and little swift.

Waders have included black winged pratincole, Caspian and white tailed plover, painted snipe, lapwing, dotterel and lesser sandplover. The raptors have included a collection in 1991 when golden, spotted, steppe and Bonelli's eagles stayed around the area for a whole day with booted, spotted, short toed, steppe and Imperial using the area in two days in October 1997. A rare record of black shouldered kite seems not out of place when you are talking about Egypt, while bateleur and white tailed sea eagle are more like rarities.

The white storks have been joined by yellow billed stork, pink backed pelican and pygmy cormorant while bean goose seems a long way away from its normal wintering grounds, shown in ancient art from Meydum! The site itself is facing an uncertain future due to development and it is hoped that it will be saved for future migrants.

The hotels around here offer yet more cover with the Clifftop Hotels, Naama Bay, Coral Bay and Shark's Bay all offering cover with several birds to their name including Hume's yellow browed warbler, African collared and namaqua dove. The walk from the Clifftop hotels back to the sewage farm has added both red tailed and Cyprus pied wheatear along with trumpeter finch, Cyprus and desert warblers, and Egyptian nightjar. Development has changed this area considerably.

The Naama Bay hotels have had dusky warbler, red throated and Richard's pipits, barn owl and western reef herons along the shore. Black bushchat was found just outside this area in 1997 while the Sinai Wildlife Centre, who look after injured wildlife at Coral Bay, had rufous turtle dove drop in to see them. This is also a good place to see a selection of gulls which come in to feed with the white storks each day. The new golf course behond Coral Bay has already had a few good birds with a male namaqua dove, roller, black necked grebe and large numbers of pipits and wagtails with new birds just around the corner.

The nearby desert can have cream coloured courser, fan tailed raven, bar tailed desert lark (record count of 108 by the airport), mourning and hooded wheatear and an old record of black crowned finch lark. Wintering birds can include Bonelli's eagle, desert finch, Cyprus and desert warbler.

The lighthouse is a known sea-watching site with passage including Leach's petrel, sooty shearwater and brown booby. Perhaps not enough time is spent here with such amazing sea-bird records from Eilat! Sooty falcon has been seen here along with hobby and red footed falcon. In recent years the local rubbish tip has added spotted eagle, Cyprus warbler and white breasted kingfisher.

**Accommodation** is widespread with 5 five-star hotels, 6 four-star, 8 three-star and 16 standard grade hotels. Like Hurghada the hotels are starting to be spread out and you may end up several kilometres from Sharm itself. Flights can be direct from many airports around the world and from Cairo, domestic flights are also available. In summer you can also fly to Hurghada if you do not want to use the ferry. Buses can take 7 hours from Cairo which is 466 kilometres, with many local departures to many other sites in Sinai. There are many taxis and also private car hire which can be arranged through your hotel or the offices in town.

Travelling from the Sharm area offers the chance to visit many sites, the **Ras Mohammed National Park** being one of the best. The area covers 480 square kilometres and even stretches around the development at Sharm, Naama and Coral Bay right up to the boundary of Nabq, as well as taking in Tiran Island. The tip of Sinai is only 30 kilometres from Sharm. Here the desert atmosphere runs down to the sea, with vast dunes and open rock. Many visitors come here to snorkel as the wildlife is as rich under the water as above and some days you may wish that you had stayed under as the heat can be unbearable!

Migration can be good down here with recent records suggesting that many birds have been missed in the past. Up to 670 Levant sparrowhawks and 464 steppe buzzards have been counted here along with white pelicans, white storks, cranes and herons mainly in the autumn. Passerines are limited by the heat and the lack of cover, with the small mangrove area really the only green area for them.

Resident western reef and striated herons breed here as well as several pairs of ospreys which like to use the pylons in the area. Both blackstart and hooded wheatear can be found here and, through the heat, even sand partridge. This should be the main sea-watching point in the country but few birders tend to use it. Records include Persian and sooty shearwater, Wilson's petrel, red billed tropicbird, brown booby and several terns including bridal, Saunder's, little and white cheeked terns. Misses come in the shape of kittiwake and skuas in spring, sooty tern in July and August, and several shearwaters and petrels when you see what has turned up at Eilat.

Tiran Island has a number of breeding birds including suspected crested and lesser crested, white cheeked and Caspian terns, white eyed gull and Kentish plover. Crab plover has been found here along with Arctic tern but the island does not get much coverage, especially for sea-watching.

A new visitors' centre is about to open at Ras Mohammed and with a bit of planting could start to bring in the passerines. An underwater walkway is planned to allow viewing of the fish and coral without getting wet. The large number of diving schools in the area offers many chances to explore more of the sea around here but it could also be a way of finding some of these rarer passing sea-birds.

To the north of Sharm is **Nabq** which is another national nature reserve where the vegetation is much more noticeable. This is found 35 kilometres from Sharm and covers 600 square kilometres. Here there are some very large areas of mangrove with many salt-loving species moving back towards the mountains - 134 species of plants in total.

The numerous bays around the mangroves attract many migrating birds especially spoonbills, white storks and pelicans. Breeding birds are similar to those of Ras Mohammed with western reef and striated heron and osprey. Migration is rarely covered and recent records of isabelline shrike from Wadi Kid on 6$^{th}$ March 1996 shows that good birds can turn up here. Waders include both greater and lesser sandplover, curlew, greenshank, Kentish and grey plover.

The diving is amazing, with old wrecked ships and graded sloping shores on a sandy bottom. The area is well worth a visit and a round trip, using a taxi, can be done - this will take the sandy coast road past the mine-fields onto the reserve and back via Wadi Kid and the main Sharm to Nuweiba road. You need to take all your water, food and diving gear as there are no facilities on-site. A visitor centre is planned but is unlikely to sell food or water.

The other site which has several coach tours a day is **St Catherine's Monastery** which is 250 kilometres from Sharm and 394 kilometres from Cairo. The monastery is set in a protected area within a 50 kilometre radius of here. The monastery's gardens and trees here provide the cover for many local and migrant birds but the number of cats around the place must remove many records before they are found!

The monastery and the surrounding hills can really get you into the Sinai specials, with birds like Tristram's grackle, scrub warbler, blackstart, Sinai rosefinch, fan tailed raven, chuckar, sand partridge, Hume's tawny owl and house bunting. The site has also had its fair share of migrants. Records of scarlet rosefinch, cinnamon breasted, little and yellow breasted buntings, rosy coloured starling, red breasted flycatcher, barred and thick billed warbler have all been found here as well as many commoner species.

New breeding birds include orange tufted sunbird and alpine swift in the mountains, with past records of lammergeyer and Verreux's eagle. Bonelli's eagle may well be still breeding in the area and lapped faced vulture may return from the captive breeding programme in Israel. Brown throated sand martin was seen here in 1985 and Eleonora's falcon in 1990.

Moving up to Mount Sinai at 2,285 metres there can be an impressive movement of raptors with eagles, buzzards and kites, but spring records of both Levant sparrowhawk and honey buzzards are worth trying for in late April and early May. To really do justice to this site it may well be worth staying over, either in the monastery's own **accommodation**, or in one of the five hotels down the road.

Siwa Oasis

Siwa Oasis

Bird catching at Marsa Matruh

Western Desert

Wood Sandpiper

Wryneck

Roller

Red backed Shrike

Crested Lark

Short toed Lark

Northern Wheatear

Moustached Warbler

Icterine Warbler

Willow Warbler

Ortolan Bunting

Corn Bunting

Lake Bardawil (Green turtle)

Greco/Roman site on Lake Bardawil

El Arish sand dunes

Purple Heron (Adult)

Greater Flamingo

Avocet

Grey Plover

Temminck's Stint

Little Stint

Curlew Sandpiper

Cretzschmar's Bunting (Male)

Cretzschmar's Bunting(Female)

Suez Park/Bay

Raptor migration

Red footed Falcon

Greater sandplover

Kingfisher

Black winged Stilt

Yellow Wagtail
(black headed)

Movenpick
El Gouna

Gifton Islands

Osprey nest/Pipe Fish

Gifton Island

Hurghada rubbish Tip

Gebel Zeit

Raptor viewpoint/St Paul's Monastry

You also have to remember that if you have come from the coast in winter the temperatures here can be extremely cold. There are buses connected directly with Cairo or with the rest of Sinai and there are even taxis on site. A small airport is here but flights were suspended (1997) and it is best to check well in advance whether it has re-opened.

Travelling west from here is Wadi Ferian which has tremendous cover before falling out onto the desert plain of the Gulf of Suez. Several records of Hume's tawny owl have come from here with records also from Wadi Mukattab and Wadi Serbal, as well as birds like Tristram's grackle and Sinai rosefinch. Also many migrants include lesser kestrel, golden oriole, bee-eater, red backed shrike, chaffinch and brambling. The find of a weaver's nest in Wadi Seih off this wadi on 1st January 1995 was thought to be the work of a male Ruppell's weaver.

The other ports of call could be **Dahab, Nuweiba and Taba**, all with expanding hotel complexes which can draw in the birds with such records as black bushchat and citrine wagtail at Nuweiba, with Wadi Watir for Hume's tawny owl. House bunting, Arabian babbler, yellow browed, Arabian and Menetries warbler, orange tufted sunbird and black capped bulbul feature at Taba along with the potential of adding some good sea birds with shy albatross already on the list!

Taba's local sewage works can have Lichtenstein's sandgrouse coming to drink and Wadi Taba has had striated scops owl wintering, while nearby Naqb has had houbara bustard. The beautiful habitat of Abu Galum which is 400 square kilometres of protected area north of Dahab is opening up for tourists, with a new visitor centre to be opened in the north of the area. Birds can be a bit thin on the ground here but the many plants, ibex and gazelles can make up for that. A new protected area is proposed for the Taba area.

The raptor passage passes through the mountains here and large numbers were formerly counted at Fjord which is 15 kilometres south of Taba. Other interesting species seen here are Verreux's eagle and lapped faced vulture. Local species here include blackstart, white crowned black wheatear and even desert warbler has been found.

There is a ferry to Aqaba from Nuweiba with two departures a day and this could be used for improving your sea-watching, especially in the summer if you want a chance to try and see some of these rare shearwaters. Taba is also a crossing point for getting to Eilat which is only 15 kilometres from the town.

**Accommodation** here consists of 3 hotels at Taba, 10 at Nuweiba and and 9 at Dahab with many more proposed along this 125 kilometre stretch of the Gulf of Aqaba. A small airport is found at Naqb just north of Taba and is run by Sinai Airways with flights every week but a new airport is planned for Nuweiba in the future which will have direct flights from many of the European capitals. Buses are regular but it is a long haul to Cairo. Taxis can be found around the hotels.

Egypt is an ever-changing country and that includes the bird sites as well - new information will be found as many more birdwatchers come to visit this fascinating country. Hopefully you will be one of them!

# Birds of Egypt

**This is a systematic list of many birds found in Egypt with information about others which have been found close by. Place names are often followed by a nearby town/city to help find your away around Egypt. Up to 516 species are documented here.**

**Ostrich** *Struthio camelus* This is a local breeding resident in the south eastern part of the country and the only breeding site in the Western Palearctic. An adult female attacked a vehicle belonging to a government agency in 1996 along the Red Sea while a family group consisting of 18 birds was seen that same year. Five birds including 2 immatures were seen in Wadi el Allaqi near Lake Nasser on January 17th 1990.

**Red throated Diver** *Gavia stellata* A rare winter visitor to the Mediterranean coast and the Gulf of Aqaba. The only recent records come from Eilat with a single bird on April 8th 1986. Two unidentified divers of either this species or black throated were seen off Lake Burullus on 12th January 1990.

**Black throated Diver** *Gavia arctica* A rare winter visitor to the Mediterranean coast and the Gulf of Aqaba. The only recent records come from Eilat with a single bird wintering there from November 10th 1987 and last seen on 16th March 1988 and one in the winter 1996/1997. [See also red throated diver.]

**Little Grebe** *Tachybaptus ruficollis* A common breeding resident in both the Delta and the Western Desert Oases where the African form 'capensis' is found. Winter records show birds in the Upper Nile as far as Aswan and even records from Abu Simbel [26th March 1994] and odd records from Lake Bardawil and Sharm el Sheikh. A count of 300 birds comes from Dakhla Oasis on 16th January 1990.

**Great crested Grebe** *Podiceps cristatus* A former breeding bird of the Nile Delta with a single pair here in the 1930s on Lake Manzala. Large concentrations on Lake Qarun with up to 3,500 in the past (only 858 birds having been counted on Lake Qarun in 1990) and small numbers wintering along the Nile to Aswan [Lake Nasser?] with 2 at Dakhla on 6th February 1998.

**Red necked Grebe** *Podiceps grisegena* A rare winter visitor with all previous/recent records coming from the Nile with two on 13th November 1977 near Beni Suef and four south of Esna on 2nd January 1986 with a single record from Sharm el Sheikh on 18th September 1980. One bird was seen at Eilat on 26th and 27th October 1986 and there are only two other records from Israel, both from the Mediterranean.

**Black necked Grebe** *Podiceps nigricollis* A common winter visitor to several large lakes in Lower Egypt with a maximum of 10,000+ counted on Lake Qarun in the 1980s [3,516 here on 21st January 1990] and small numbers on Wadi Natrun [200+], Lake Bardawil [200+], El Malaha [1000+], Bitter Lakes [400+]. The largest concentration counted in recent times numbered 5,811, wintering on the southern half of Lake Nasser during 1994/95. [Birds formerly from Lake Qarun!] Small numbers are seen elsewhere with the Nile Valley producing a maximum of 40+ in the duck flocks between the Esna/Edfu section. Other records are from Dakhla Oasis, Wadi el Rayan, the Gulf of Suez and Aqaba.

**Shy Albatross** *Diomedea cauta* The only Western Palearctic record from 20th February to 7th March 1981 at Eilat, later found dead. A sighting also came from Taba proving that all these sea birds found at Eilat may become Egyptian records if there was someone there to record them.

**Schlegel's Petrel** *Pterodroma incerta* Two records both in the1980s and from Eilat with one being photographed during its stay [18th-24th April 1989].

**Soft plumaged Petrel** *Pterodroma Mollis* One off Eilat on 25th March 1997.

**Cory's Shearwater** *Calonectris diomedea* The Mediterranean population probably use the Egyptian coast a lot more than records show. The wardened reserve of Zaraniq [Lake Bardawil] shows many more records than anywhere else but, with promontories like Ras el Hikma rarely used, many records go astray. Birds are now regularly seen in both Gulfs with Eilat featuring high, with up to 300 in July off the coast. Hurghada has had some recent records in the 1990s.

**Streaked Shearwater** *Calonectris leucomelas* The summer gathering of shearwaters from the beach at Eilat included around 1-3 individuals 3 kilometres offshore in 1992 and 1993 and further records came from pelagic trips in Egyptian waters.

**Pale footed Shearwater** *Puffinus carneipes* A single record again from Eilat on the 15th August 1980.

**Sooty Shearwater** *Puffinus griseus* There are few records from the Mediterranean coast but also few birdwatchers! Zaraniq has produced the bulk of the records. On the other hand this bird can be seen for 8 months of the year from Eilat with only a handful of records from the Sinai coast! The maximum at Eilat is 43 in May.

**Yelkouan Shearwater** *Puffinus yelkouan* Sea-watching produced 16 at El Arish on the 16th January 1995. This is yet another unrecorded species along the Mediterranean coast. Up to 41 birds were found dead along the beach at Lake Manzala in March 1990.

**Wedge tailed Shearwater** *Puffinus pacificus* Three records starting in November 1983 from Quiser, March 1988 from Port Said and from Hurghada during mid-June 1992 when 6 were reported, have all referred to this species. No records have yet been accepted from Eilat where streaked shearwater has been suggested for this species.

**Little Shearwater** *Puffinus assimilis* No records so far for the country with only one record from the Mediterranean for Israel.

**Persian Shearwater** *Puffinus persicus* The first record came in 1991 during September from Ras Mohammed while other records are from pelagics off this area.

**Mascarene Shearwater** *Puffinus atrodorsalis* A single bird observed from 18th - 21st June 1992 off Eilat shows how little is known about an area and what can be found. This is a newly-described species and sets the scene for more observers to patrol these coastal waters.

**Wilson's Storm Petrel** *Oceanites oceanicus* No firm record for this species in recent times although petrels thought to be this species have been seen around Ras Mohammed, Brother's Reef and to the south of the border with Sudan. Only a single record comes from Eilat on 1st June 1983.

**European Storm Petrel** *Hydrobates pelagicus* No records for this species in Egypt but records are found for Israel in the Mediterranean.

**Leach's Storm Petrel** *Oceanodroma leucorhoa* This is another species which is under-recorded with several records occurring in the Mediterranean off Israel and Gaza. A freshly-dead bird was found inland in a bird net on 25th April 1990 and 32 were seen off Zaraniq on 25th March 1997. Only one record from near Sharm el Sheikh in August 1978

**Madeiran Petrel** *Oceeanodroma castro* A single bird at Eilat on 6th July 1983.

**Red-billed Tropicbird** *Phaethon aethereus* Ras Mohammed is the main sea-watching site for this species with most records coming during March to September although there was a single record from Hurghada on 27th January 1990. One seen in April and November off the north beach, Eilat 1991 and two off Gifton Island, Hurghada on 2nd September 1995. It is a former breeding bird from the islands off Hurghada [eight islands presumed used in 1983!] and now only thought to breed in the southern islands of the Egyptian Red Sea.

**Gannet** *Sula bassana* Another unrecorded species with several old records in the 1980s and an immature flying off the coast 30 kilometres west of Alexandria on 28th February 1997 and one off Zaraniq on 9th May 1997.

**Brown Booby** *Sula leucogaster* A breeding bird of the Red Sea coast with around 50+ pairs breeding. Only one site known in the southern Egyptian Red Sea. Non-breeding birds seen regularly at Eilat from November to mid-March and birds seen from Ras Mohammed i.e. one on 2nd February 1997. There is a single record from the Mediterranean in January 1992 near Haifa, Israel.

**Great Cormorant** *Phalacrocorax carbo* An increasing number of birds thought to come from Central Europe and Russia using the Mediterranean lakes in winter with up to 28,522 counted in 1990. Damage is being reported to fish stocks and control may be used now before the reason for this increase is understood. The species is also increasing in nearby countries. Wintering birds can also be found along the Nile and even Dakhla Oasis. A day's count while cruising during winter can total 170 birds.

**Pygmy Cormorant** *Phalacrocorax pygmeus* A single bird was found exhausted at Sharm el Sheikh on 2nd January 1982. This was the same year as Cyprus had its first record. The nearest breeding colony is now in Israel with single winter records from Eilat in November 1986 and February 1989.

**Long tailed Cormorant** *Phalacrocrax africanus* A former breeding bird especially on Lake Qarun. The first returning birds moving north into Lake Nasser started in 1988 with the second record on 23rd-24th October 1991 with a maximum of 25 in March 1993, seven on 27th October 1996 and 14 in late November 1996, south of Abu Simbel. The peak time to see this bird in Egypt is recorded as the summer but, due to the temperatures and the sheer size of Lake Nasser, little information exists on other localities in the lake. Cruise boats on the lake may well increase the records in years to come.

**Shag** *Phalacrocorax aristotelis* Birds found off the Mediterranean coast are now believed to come from breeding stock found off an island 10 kilometres off Ras el Hikma. This is the first known breeding site in the country. A single bird was seen close by at Fuka. A single bird was here on 3rd March 1995 and 6 birds were seen on an island near Marsa Matruh on 24th May 1997. A single record comes from the Red Sea off Hurghada in November 1982.

**Darter** *Anhinga rufa* A former wintering bird probably from the Turkish population which died off due to their breeding area being drained in the 1950s. The only recent records come from 3 sightings in 1985 and in 1986 near Edfu.

**White Pelican** *Pelecanus onocrotalus* A common autumn and spring migrant with records indicating a Sinai, Red Sea, Eastern Desert, Nile [Qena] crossing of the country with up to 1,157 wintering on the southern half of Lake Nasser during 1994/95. Abu Simbel itself held 600+ in January 1997 and, given the size of the lake, many more go uncounted.

**Dalmatian Pelican** *Pelicanus crispus* Very few records exist of this species with a scattering of records in the 1970s and 1980s with 3 at Sohag seen from a Nile cruise on 10th November 1977 and 7 found on El Malaha in the winter of 1979/80. A single bird wintered at Eilat in 1982/83.

**Pink backed Pelican** *Pelecanus rufescens* Abu Simbel is the best site to see this species with a maximum of 95 there in late April 1994 and 33 were there on 28th April 1997. Some records come from the Nile itself with 2 at Kom Ombo on 21st May 1993 and between the dams at Philae with 2 there on 28th September 1990 and 2 again on 30th August 1993. A single bird was at Sharm el Sheikh on 24th December 1995.

**Bittern** *Botaurus stellaris* Formerly thought to be a rare winter visitor but the Nile birdwatching cruises have added many records of this shy species, especially from Qena to Kom Ombo. With very little reed found here birds have been seen as far south as Abel Simbel with records on 8th December 1995 and again on 25th January 1997. A single record comes from Hurghada on Abu Mingar Island on 31st December 1993 and a single bird was seen at Wadi el Rayan on 28th January 1995 and again on the 10th February. Another record of a single bird came from a small marsh on the Bitter Lakes on 1st April 1995. A summering record comes from 35 kilometres north of Zafarana on 19th June 1990.

**Little Bittern** *Ixobrychus minutus* A common breeding bird of the Nile Delta with 1,000+ pairs. Possible breeding sites have also been suggested along the Nile with birds seen in May. It is a common migrant with some years' numbers exceeding 2,000, moving inland from Zaraniq, and wintering numbers make it an easy bird to see on a cruise along the river.

**Night Heron** *Nycticorax nycticorax* A local breeding bird of the Nile nesting on islands from Qena to Aswan with a new colony found 20 kilometres north east of Bilbeis in the Delta with 150+ nests. Large numbers winter especially in the Nile Valley. Migrants were seen moving north from Abu Simbel on 9th April 1997. Migrants move on a broad front across Egypt with up to 20 wintering at Dakhla Oasis 1997/98.

**Striated Heron** *Ardeola striatus* Small numbers nest on the islands along the Red Sea coast. Breeding birds also seen along the Nile from Kom Ombo and Gezira Ghazal, Aswan in 1990. Breeding is also recorded along the South Sinai coast with juveniles seen at Nabq and Ras Mohammed. Recent records also show a movement north up the Nile as far as Nag Hamadi, with 2 just south of here on 30th September 1992. Three recent records from Abu Simbel with the first in 1995 and one on 9th April 1997.

**Squacco Heron** *Ardeola ralloides* An uncommon breeding bird of the Nile Valley with the best observed colony on Kitchener's Island, and Gezira Ghazel, Aswan. Up to 5 pairs now breed on an island near Abu Simbel. The Delta has a large colony of 500+ pairs near Damietta and one of 100+ pairs near Bilbeis. A common migrant passing, in often thousands, from sites such as Zaraniq, North Sinai. Smaller numbers winter along the Nile with birdwatching cruises expecting at least 50+ birds a day and some counting several hundred.

**Cattle Egret** *Bubulcus ibis* The commonest breeding heron found in Egypt but surprisingly, no records from Ancient Egypt! Best observed colonies at Giza Zoo, Cairo [7,269 in 1990] and increasing at Kitchener's Island, Aswan [400+] with a new colony on Lake Nasser off Abu Simbel. The largest colony in the Delta is near Bilbeis where up to 5,000 pairs are found. Fewer birds found along the Red Sea [313 nests at Suez in 1990] but domestic stock and rubbish tips like Hurghada can always turn up some birds.

**Black Heron** *Hydranassa ardesiaca* Singles at Eilat 1982 and Yemen 1996.

**Western Reef Heron** *Egretta gularis schistacea* Found breeding and wintering along the Red Sea coast with smaller numbers around the southern end of Sinai. Dark morph birds found more commonly towards the Sudanese border. The tourist industry has not scared the birds away from sites like Hurghada and Sharm el Sheikh!

**Little Egret** *Egretta garzetta* A local breeding bird of mainly the Upper Nile with the best-viewed colony on Kitchener's Island, Aswan [1,000+] and a new colony off Abu Simbel. The Delta has colonies at Lake Manzalla near Damietta and near Bilbeis. Birdwatching cruises have indicated increased numbers from Luxor to Aswan [500+ birds] during the summer months while migration and winter can show as many as 1,038 birds between Esna and Kom Ombo [September 1993].

**Great white Egret** *Egretta alba* Fewer birds seen than any other heron during Nile cruises. A maximum of 49 seen in March 1987 was the best with some cruises even failing to see them at all during spring or autumn cruising. The Delta lakes seem to hold a winter population of up to 1,000 birds. Migration seems to start early with most birds departed from Lake Manzala by the end of March. During 1990, 691 were counted in the first week, dropping to 214 and 137 by the third week. A count of 23 was found in the last week of April. There were 2 birds at Siwa Oasis on 27th September 1997 and 2 at Dakhla Oasis on 6th February 1998.

**Grey Heron** *Ardea cinerea* A common migrant and winter visitor to most parts of the country with 230 counted at Dakhla Oasis on 6th February 1998. Ringing returns show wintering birds travel from as far as Poland and Russia.

**Black headed Heron** *Ardea melanocephala* Two were found 8 kilometres north of Aswan on 13th December 1995. An immature bird was at Eilat from 19th October to 15th December 1987.

**Purple Heron** *Ardea purpurea* Common along the Nile and can be found during any month of the year but no records of breeding yet. The best count on a cruise came from September 1989 when 139 were counted in 8 days. Good numbers recorded also in winter for what is usually a fairly secretive bird. Large counts have been obtained on passage from Zaraniq with up to 2,000 birds and Suez with flocks of up to 59 birds.

**Goliath Heron** *Ardea goliath* Small numbers now breed along the southern Red Sea coast and not too surprisingly several records are found now beyond the breeding area. Two at Suez on 8th April 1990, one on Abu Mingar Island off Hurghada on 31st December 1993 and again on 2nd September 1995. A single bird in mangroves 36 kilometres south of Safaga on 6th May 1994, a single adult was seen 10 kilometres south of Safaga on 25th September 1995 and a juvenile 5 kilometres south of Quiser. The Nile has had several records as well, with one at Aswan High Dam on 11th March 1992 with also a juvenile at Abu Simbel. A single bird was 8 kilometres north of Aswan on 13th December 1995 and the farthest north up the Nile was from Esna on 17th January 1993.

**Yellow billed Storks** *Mycteria Ibis* One of the many attractions from Abu Simbel recently with the largest count here being 89 on 1st-3rd May 1994 and 36 here on 29th April 1997. A juvenile was present at Sharm el Sheikh from 13th August to 3rd September 1995 and not to be beaten, the birdwatching cruises spotted 9 flying south on 26th May 1993 just south of Balliana. Aswan has also recorded single birds. Birds have now been seen as far north as Turkey.

**Black Stork** *Ciconia nigra* With 10 - 20,000 birds passing through Israel the records are very low for Egypt. The main flocks numbering 100s are watched from Suez south with visible migration seen from Ain Sukhna, St. Paul's Monastery and even Hurghada. Small parties are seen along the Nile with an exceptional record of 100+ near Edfu in October 1990, most being individuals or 5-10s, as over the Valley of the Kings in 1986 and 1992. Odd birds can stay for the winter or summer, especially along the Nile. Two satellite-tracked birds moved from the Czech Republic to winter in northern Ethiopia. One bird was killed there while the other bird returned again passing through Egyptian air space.

**White Stork** *Ciconia ciconia* A large passage takes place through Egypt during the spring and autumn migration with small numbers wintering and very small numbers summering. Records occur from around the country but the main migration can be seen from South Sinai to Suez south through the Qena gap on to the Nile and south past Abu Simbel. Half a million birds have been estimated to make the crossing [85% of the world's population]. Type 'stork migration map' on the Internet and follow their journey. Ancient sites, visited during the Nile cruises, such as Philae had 15,000+ in 1986. Karnak has added records like 2,700 in an hour, crossing the columns in 1993 and Edfu also had 5,100 in 1993.

**Glossy Ibis** *Plegadis falcinellus* Mainly a passage bird with increasing numbers wintering in the Upper Nile and small numbers summering with breeding suspected near Bilbeis in the Delta. This is also a new breeding species for Israel. A common bird seen on most of the birdwatching cruises with totals numbering, during the several days cruising, 221 birds in 1986, 189 in 1992 and 456 in 1993.

**Bald Ibis** *Geronticus eremita* With a declining world population no recent records exist of a bird well used here in ancient art The only nearby record comes from Israel with 3 flying north in the Arava Valley on 8th April 1995

**Sacred Ibis** *Threskiornis aethiopicus* Another former breeding bird of the Nile Delta with few records in recent times. Birds seen in Sinai during 1886 and 1891 seem to be the last records but an amazing record comes from a summer birdwatching cruise with a single bird seen with both binoculars and telescope on 20th June 1986. There was also another record from Agilkia Island, Aswan of an adult on 3rd May 1990, so hopefully more records will come in the future from the nearby Sudanese breeding grounds!

**Spoonbill** *Platalea leucorodia* A breeding bird of the Red Sea coast with a small number of pairs using the mangroves to nest in. A common migrant with a smaller number wintering along the Nile and Lake Nasser. Philae is the best ancient site to see this species with birds migrating through or feeding in the shallows near the village of Shallal. Birds are also commonly seen off the cruise ships on the Nile with total figures of a cruise being 259 in 1984, rising to 497 in 1993.

**Greater Flamingo** *Phoenicopterus ruber roseus* A breeding bird of El Malaha in North Sinai with summering birds also on Lake Bardawil with artificial nest sites made for the birds here in 1996 but as yet, no proof of breeding. Wintering birds can be found in the Delta Lakes with up to 20,000 birds often present while migration has given records along the Upper Nile and Lake Nasser where wintering birds can also be found. Nile records include 43 flying north on 24th May 1993 at Luxor and 110 flying north on 26th August 1993 at Dendera.

**Lesser Flamingo** *Phoenicopterus minor* First record for the country and Western Palearctic came on 27th November 1992 from off the cultivated land north of Abu Simbel.

**Fulvous Whistling Duck** *Dendrocygna bicolor* A single bird at Eilat on 30th March 1993 of a species found in Central Africa but some are also kept in captivity.

**Mute Swan** *Cygnus olor* A rare bird in Egypt with an influx in the winter of 1984/85 and 1985/86 with a maximum of 10 birds in the Delta and mainly singles in the Lower Nile Valley.

**Whooper Swan** *Cygnus cygnus* A former winter visitor to the Delta region with no recent records other than a single bird seen along the Nile Valley in February 1976 near Abu Tig. Three birds were seen at Eilat on 17th and 18th March 1985 and Cyprus had 2-3 birds wintering in 1994/95.

**Bean Goose** *Anser fabalis* A bird of ancient times as seen in the 'Geese of Meydum'. Two birds were photographed at Sharm el Sheikh in October 1982.

**White fronted Geese** *Anser albifrons* A former winter visitor around the Delta [Geese of Meydum] with odd records now along the Nile and even as far as Abu Simbel during the 1980s. Several have been recorded on birdwatching cruises along the Nile. A single goose flying with Egyptian geese at Abu Simbel could have been this species in January 1997. A party of 6 wintered at Eilat from November 1993 to February 1994.

**Lesser White fronted Geese** *Anser erythropus* Two adults wintered along with the white fronted geese at Eilat during November 1993 and February 1994. No recent records in Egypt.

**Greylag Goose** *Anser anser* Like all the grey geese few make it this far to winter nowadays with only 4 records in total, three of these coming from the Sinai during the 1970s and 1980s.

**Barnacle Goose** *Branta leucopsis* Only one record exists of this species in Egypt and that was from April 1931.

**Brent Goose** *Branta bernicla* No recent records exist.

**Red breasted Goose** *Branta ruficollis* No recent records exist in Egypt with the fabulous portrait in the Cairo Museum being the only one to remain for so long, but the last record in Israel comes from 1984.

**Egyptian Goose** *Alopchen aegyptiacus* A common breeding bird of Lake Nasser with previous breeding sites along the Nile, abandoned after the making of the High Dam. A winter visitor to many parts of the Nile with El Til being the farthest north the species was seen on birdwatching cruises other than one record from outside Cairo in 1985.

**Ruddy Shelduck** *Tadorna ferruginea* A declining number of records from the Delta with few birds seen on the cruises other than 2 flying north on 21st August 1993 near Beni Suef. Other records have been a single bird found at Zaraniq on 12th October 1994, a single bird on the River Nile at Esna on 30th September 1996, up to six birds seen south of Kom Ombo on 29th December 1996, matching with an influx at Eilat with 28 on 15th November 1996.

**Shelduck** *Tadorna tadorna* This bird is found along the northern coast in winter with numbers fluctuating from year to year with 2,467 counted in 1990, with 1,350 of these birds from El Malaha. Few records from elsewhere with Kom Ombo the farthest south that birds have been seen.

**Spur winged Goose** *Plectropterus gaminesis* This bird is found in feral flocks of ducks and geese in the Delta area but it is not known if any wild birds have really occurred in Egypt. Bones have been found from historical times but these too could have come from domestic geese.

**European Wigeon** *Anas penelope* A common winter visitor with several thousand found in many sites around the Delta with 3,000 on Lake Idku, 19,000 on Lake Burullus with hundreds along the Nile with peaks of 2,000 between Beni Suef and El Til and large rafts amongst the other duck at Edfu. There seems to be many birds now wintering on Lake Nasser but so far only 5,000+ counted in 2 flocks in February 1998.

Egyptian Vulture

Steppe Eagle

Collared Pratincole

Green Desert

Gebel Elba

Wadi Abu Ziran [Bir Beida]

The mangrove south of Safaga

Dark morph of Reef Heron

Osprey

White eyed Gull

Caspian Tern

Sooty Gull

Children of the Island

Karnak

Wetlands

Thebes

Queen Hatshepsut's Temple

Bimaculated Lark

Red throated Pipit

Bonelli's Warbler

Crocodile Island

Spanish Sparrow

Movenpick Hotel+Silk cotton tree+flower

Luxor – ruins at Karnak and the Brown-necked Raven resting in the temple

Brown necked Raven-Karnak

The Nile-Little Egret, Night Heron, Gull billed Tern, Garganey, Cattle Egret, Striated Heron

**Gadwall** *Anas strepera* A wintering bird of the Delta Lakes with Lake Idku having 200 in 1990 and 838 along the Nile Valley. Most cruises were lucky to see 10+ birds on their trip!

**Teal** *Anas crecca* A passage and wintering species with large numbers around the Delta Lakes including 2,000 on Lake Burullus, 4,400 at Lake Manzala and 3,000 at Lake Qarun. Counts often were above 1,000 birds along the Nile with 2,100 between Tima and Balliana.

**Cotton Teal** *Nettapus Coromandelianus* A female was at Aquaba, Jordan on the 9th April 1997.

**Mallard** *Anas platyrhynchos* A winter visitor with up to 5,000 counted mainly around the Delta Lakes with smaller numbers of 50+ seen along the Nile.

**Pintail** *Anas acuta* A common passage and winter visitor with birds found at many inland sites such as Lake Qarun and Wadi el Rayan. Common on the lower and middle Nile where counts have exceeded 500 birds. Late spring records include a pair on 31st May 1990 at Edfu on the Nile.

**Garganey** *Anas querquedula* A very common passage duck with exceptional numbers counted some years of Zaraniq with a maximum of 112,883 counted over several days in September 1986 [221,661 in 1981]. Much smaller numbers are spread along the Nile and counts of 884 have been made on the birdwatching cruises. Up to 5 birds were found as late as 24th May 1990 near New Alexandria.

**Blue winged Teal** *Anas discors* The first record for Egypt was of a full plumaged drake found for sale in a Port Said market on 11th March 1990. The bird had been shot the previous week on Lake Burullus. This would also be the 12th record for the African continent.

**Shoveler** *Anas clypeata* A very common winter visitor and passage migrant with large numbers in the Delta Lakes, Nile Valley and Lake Nasser. Many parties of 500+ were seen from Nile cruises between November and March. Lake Burullus was the single most important wintering site for this species in the Eastern Mediterranean and Black Sea region with a maximum of 63,500 counted here. This had fallen to 15,427 in 1990. The total count for the whole country reached only 45,140 that same year.

**Marbled Teal** *Marmaronetta angustirostris* A very rare breeding bird with a small wintering population especially along the Nile. Several records of birds in winter come from the Edfu-Esna stretch where mixed duck numbers can reach several thousand. The last record from here was on 7th December 1991. Two birds were found on the north coast at Alexandria on 22nd May 1996 and a single bird was seen from Abu Simbel on 24th January 1997.

**Red crested Pochard** *Netta rufina* A winter visitor to the Delta and the Nile in small numbers with birds having been seen from the cruises especially around Esna and Edfu. Other records come from the Western Desert with a drake seen at Wadi el Rayan on 28th January 1995 and Lake Qarun held 8 on 17th February 1995.

**Pochard** *Aythya ferina* A common winter visitor to the Delta area with Lake Burullus holding 7,357 in 1990 and the Western Desert lakes such as 2,637 on Lake Qarun in 1990 and 1,825 feeding with coot on Wadi el Rayan the same year.

**Ferruginous Duck** *Aythya nyroca* A fairly common winter visitor with birds found on the Delta with Lake Burullus being the main area for the species [with 576 there in 1990] in the Eastern Mediterranean and Black Sea area. Other birds can be found along the Nile Valley especially between Esna and Edfu. A large concentration of 130 were found between the old and new dams at Aswan on 5th February 1995 with 4 at Abu Simbel on 25th March 1995. A gathering of 730 were present between Dendera and Edfu on 3rd March 1997. Away from these areas small numbers were found on Lake Qarun and on the Red Sea coast at Hurghada rubbish tip in 1997.

**Tufted Duck** *Aythya fuligula* A common winter visitor with large numbers found on Lake Burullus, Lake Qarun and Lake Nasser. The largest wintering concentrations for the country were 19,281 on the southern part of Lake Nasser during 1994/95. Only 10,609 were counted in the whole country in 1990.

**Velvet Scoter** *Melanitta fusca* Only one recent record of a female in Suez Bay during 24th February to 12th March 1982.

**Smew** *Mergus albellus* No recent records of this species other than several between 1947 and 1979 but the remains of this species were found at Kom Ombo from ancient times. Singles were at Eilat in 1968 and 1989.

**Red breasted Merganser** *Mergus serrator* An increasing number of records due to more observers especially along the Mediterranean. A flock of 19 were recorded off Ras Kanissa on 16th December 1988. Four drakes were seen on Lake Bardawil on 12th January 1990, one was at Lake Qarun on 3rd December 1994, a female along the Mediterranean in 1995, a drake at Eilat on 19th January and six birds remained at Eilat from 5th December 1988 to 4th June 1989.

**White headed Duck** *Oxyura leucocephala* With a declining world population no recent records exist of this species.

**Honey Buzzard** *Pernis apivorus* A common passage migrant with lack of full counts on migration [counts in Israel can be as high as 851,598 in spring!] Best viewpoints are along the Red Sea and Gulf of Suez with other records from the Nile from Qena south. Records from Gebel el Zeit indicate up to 3,400 on 6th May 1983 while 1,607 birds were moving across the Suez on 6th May 1994 eleven years on. Cairo even claimed 52 birds on 31st May 1990 as being one of the latest records for spring migration.

**Crested Honey Buzzard** *Pernis ptilorhynchus* The first record for this area of the Middle East is from the Eilat mountains on 14th May 1994 followed by the first for Egypt on 9th May at Bir el Ambagi. This was an adult male and only the fourth record for the Western Palearctic. Another was found on the border with Israel at Nizzana on 10th and 11th November 1996.

**Black shouldered Kite** *Elanus caeruleus* An increasing species in Egypt due to the continuous development of farmland. Found in most cultivated areas with Nile cruises seeing a maximum of 140 birds on a trip. Away from these areas two adults were in Gebel Abas Fesha on 3rd October 1993 and a juvenile was at Sharm el Sheikh in the Autumn of 1995. There are some indications of persecution by foreign gunmen especially along the Faiyum.

**Black Kite** *Milvus migrans* A breeding bird of the Nile Valley and Delta areas. Commonly seen over Cairo with display and courtship viewed in the sports complex on Gezira Island, Cairo; also display watched at Luxor and Aswan. Frequently seen at ancient sites all the way south to Abu Simbel. The most recorded roost watched is on Kitchener's Island, Aswan by Nile cruises where up to 250-300 birds can be seen most of the year. Migration hot spots include up to 3,861 over three months at Suez, with others at Hurghada, the Red Sea and Nuweiba on Sinai.

**Red Kite** *Milvus milvus* An irregular migrant mostly picked up from 'raptor counting' along the Suez and Gebel el Zeit. The most recent records was 1 at Gebel Asfar on 15th November 1991 and 2 on 9th April 1992 at Gebel el Zeit. A single record at Suez on 21st March 1998. Few records are found at Eilat.

**African Fish Eagle** *Haliaeetus vocifer* An old record was of a bird shot at Aswan in 1947 and also a record in November 1967 which was rejected by the then rarities committee. One possible bird seen in November 1997 on Lake Nasser. A bird's remains were found at Elephantine, south of Aswan, from ancient times. No records exist from Eilat.

**White tailed Eagle** *Haliaeetus pelagicus* A former breeding bird of the Delta but no other records exist other than a juvenile was found at Sharm el Sheikh from 23rd August until 24th October 1995. A very rare migrant over Eilat and a winter visitor to the north of the country where a reintroduction scheme has been running at Hula, Israel.

**Lammergeyer** *Gypaetus barbatus* A bird which has declined in most of its range in Egypt but still hangs on in the south east corner of the Red Sea and South Sinai. There are some new records of birds crossing the Israeli border in the north east corner of Sinai and one or two records of migrants, mainly along the Suez. One was seen between Luxor and Esna on 17th December 1995. A species which may increase if effort was put into using feeding stations as now in many countries, especially Spain.

**Egyptian Vulture** *Neophron percnopterus* A local resident and breeder especially along Lake Nasser and the Eastern Desert. Migrants do pass through the country with some wintering. The best sites to see this bird are at Aswan, Philae with up to 30 birds seen here, and Abu Simbel. Migrating birds can be seen from Wadi Hagul, Suez, Ain Sukhna and St. Paul's Monastery along the Red Sea with St. Catherine's a good site in Sinai.

**Griffon Vulture** *Gyps fulvus* A former breeding bird, even around Cairo, and a rare winter visitor with migrating birds declining. Seen from several sites on migration with Suez standing out. Nile birds seen, especially from Aswan and Philae. Type 'stork migration map' and follow a Griffon Vulture's movements on the Internet.

**Ruppell's Vulture** *Gyps rueppelli* A very rare bird with no recent records. Abu Simbel is the place to find one.

**Lapped faced Vulture** *Torgos tracheliotus* A rare breeding bird mainly of the south east corner of the Eastern Desert where a pair were seen at Gebel Elba in December 1997 and gone from Sinai. Records of birds mainly from 25 kilometres north of Abu Simbel with three on 31st January 1993, four there in April 1993, a pair 32 kilometres south of Aswan in May 1994 and an immature was at Bir El Ambagi near Quiser on 9th May 1996. A pair with a nest were found in March 1998 near Aswan. A captive programme is in place in the Negev Desert and future releases may well see birds back into Sinai.

**Black Vulture** *Aegypius monachus* A rare migrant with singles usually seen from Eilat heading over the border. There were two records from Nile cruises with 2 birds possibly from Philae on 9th October 1985 and a single bird also from here on 26th March 1986. A single bird was seen on 28th December 1993 eighty kilometres north of Aswan.

**Short toed Eagle** *Circaetus gallicus* A Common migrant but a rare breeding bird with few recent records. Suez is the place to see the major passage of this bird with up to 12,000 counted here between September and November. It is commonly seen on the Nile cruises mainly as singles but sometimes in small groups of 1-5. Other records come from Wadi Hagul, Ain Sukhna, St. Paul's Monastery and Hurghada.

**Bateleur** *Terathopius ecaudatus* A single pair has been known to breed at Gebel Elba in the south east corner of the country. Other records have come from Abu Simbel with a first year male in 1985, one at Sharm el Sheikh on 15th December

1988 and a juvenile at Eilat in the second half of March 1996, this being the sixth record for Israel.

**Marsh Harrier** *Circus aeruginosus* This bird is mainly a passage bird with small numbers staying to winter in the Delta Lakes and Nile Valley. Peaks of 115 birds at Lake Burullus in winter with 314 counted overall. One of the easiest birds to watch on Nile cruises due to the height of the boat and that sweeping flight over vegetation or open water. Some cruises managed to see as many as 55 birds, out of which only one male was recorded in September.

**Hen Harrier** *Circus cyaneus* A wintering bird of the Delta and Nile Valley with up to 38 counted in 1990 with 30 of these coming from the Delta. Females or 'ring tails' were seen away from here at Wadi Natrun in 1997 and as far south as Sohag on the Nile cruises. Passage has been seen along the Red Sea in small numbers from Suez.

**Pallid Harrier** *Circus macrourus* The bird that many birdwatchers dream of seeing, with passage and wintering birds found here. A good bird from the Nile cruises with birds seen mainly in March and September with records from Beni Suef to Abu Simbel. The most seen on one trip was 7 in 1992. Wintering birds are seen mainly on the Delta with records from Lake Manzala, the Port Said-Ismailiya road, Lake Bardawil and 15 were wintering in the North West Negev/Egyptian border during 1996/97.

**Montagu's Harrier** *Circus pygargus* A passage migrant which moves right across the country. Fewer records than pallid on the Nile cruises but several unidentified 'ring tails' also seen, mainly put as either pallid or Montagu's. Birds have also been seen at Wadi el Natrun, Lake Qarun, Lake Manzala, Suez, Hurghada and Bir Beida.

**Gabar Goshawk** *Micronisus gabar* No recent records exist of this species but the old records were thought to be of birds moving out of Sudan along the Nile.

**Goshawk** *Accipiter gentelis* Few records exist of this species but most are thought to have moved through the Suez Canal area. A female was seen at Sheikh Zeweid on 27th December 1994.

**Sparrowhawk** *Accipiter nisus* Small numbers winter in Egypt with most birds moving through the country as migrants. Wintering birds have been seen at Cairo, Gebel Asfar, Lake Qarun, Aswan and Abu Simbel with up to 15 counted at these various sites. Passage birds are more often found at Suez with up to 400 birds seen over a 3-month period.

**Levant Sparrowhawk** *Accipiter brevipes* A steady movement of birds seems to take place from April into May and returning from the end of August to the start of November. Highest counts in the country so far include 2,347 over Gebel Zeit on 27th April 1994 while in contrast, 25,522 were counted at Eilat on 25th April 1994. Returning birds were counted over Ras Mohammed on 5th October 1996 with 670 present, this being the highest autumn count in the country. Other sites for this species include Hurghada and Bir Beida, near Quiser along the Red Sea.

**Buzzard** *Buteo buteo* The western form of this species is that it is mainly a winter visitor with up to ten birds wintering in the Delta with other birds seen at Gebel Asfar.

**Steppe Buzzard** *Buteo buteo vulpinus* A very common passage bird with spring movements along the Red Sea often involving several thousand birds a day. Suez counts have reached 80,000 over a two and a half month period compared with Eilat's 465,827 birds during a 3-month period. Anywhere along the Red Sea coast can show movement of birds even out onto the islands off Hurghada. Commonly seen on Nile cruises in spring and autumn. Sinai also has large numbers with Mount

Sinai area a favourite site for watching, depending on the wind direction and temperatures. Zaraniq had 1,453 on 21st March 1997 and around 464 were flying over Ras Mohammed on 5th October 1996.

**Long legged Buzzard** *Buteo rufinus* An uncommon breeding bird in Egypt with most birds migrating through or wintering. Several nesting pairs are found in Sinai. Suez has recorded many birds on passage with 168 between 4th March and 21st April. Birds are recorded throughout the country in winter with several records on Nile cruises.

**Lesser spotted Eagle** *Aquila pomarina* A spring and autumn passage bird with large numbers counted from Suez where an estimated 40,000 pass through in the autumn, with much smaller numbers seen in the spring. Inland sites include Wadi Hagul where several were seen there in 1997, St. Paul's Monastery, Lake Qarun and the Nile Valley where birds have lingered into November. In 1995 radio-tracked birds were followed from Germany south across Sinai along the Eastern Desert, moving back to the Nile in central Sudan and wintering around Lake Tanganyika. The migration took between 7 and 8 weeks to complete for both directions and a narrow corridor was passed through on both trips. A total of 19,400 kilometres were covered by one wintering bird on migration.

**Spotted Eagle** *Aquila clanga* This was once a common winter visitor to Egypt but few birds winter today with records from the Nile but not the Delta. A single bird wintered around Sharm el Sheikh in 1994/95 and Dahshur and Aswan in 1997/98. A total of 117 have wintered in Northern Israel. During spring migration birds have been seen in Wadi Hagul and Suez. Radio-tracked birds have been studied but no results are available yet.

**Steppe Eagle** *Aquila nipalensis* The most common eagle on passage with up to 100,000 estimated from a maximum count of 64,880. This passage crosses Sinai down to the Red Sea and along the Eastern Desert in autumn, with smaller numbers counted in spring. More birds have been seen in the Nile valley in the autumn than in the spring with Aswan standing out as a site of passage. A winter record came from Dakhla Oasis in the Western Desert in January 1990. Follow the migration on the Internet, typing in 'stork migration map'.

**Tawny Eagle** *Aquila rapax* A single bird wintered in the north west Negev/ Egyptian border in 1996/97. This was only the second record for Israel with no recent Egyptian records accepted.

**Imperial Eagle** *Aquila heliaca* Small numbers pass through the country on passage, with more records in autumn than in spring. Highest counts have included up to 500 birds in a seven-week period. Several wintering records have come from the Nile with one wintering around the southern half of Lake Nasser during 1994/95. North west Negev held 35 wintering in 1996/97 but no birders on the Egyptian side!

**Golden Eagle** *Aquila chrysaetos* A pair was found west of Qusaima in northern Sinai during the 1990s with few other pairs located but there is a good breeding success over in the north west Negev with a high rate of young birds which may allow more birds to cross the border. Very few records exist from the rest of Egypt with odd birds seen at Suez on migration and Sharm el Sheikh where a bird was seen with one Bonelli's, one spotted and three steppes all around the trees of the sewage works in the autumn of 1991.

**Verreaux's Eagle** *Aquila verreauxii* A wide gap exists between the birds found in South Sinai and the ones around Gebel Elba. Breeding seems patchy but the birds in the north raised young, off and on, up to 1993 when a first winter bird and an adult were seen. Around the area of St Catherine's Monastery is another likely place to see this species. At least two pairs are thought to breed around Gebel Elba where a pair were seen in December 1997.

**Bonelli's Eagle** *Hieraaetus fasciatus* Two to three pairs may breed in Sinai and Gebel Elba with very small numbers seen on passage. Two birds were seen at Gebel Serbal in June 1990 and two were seen on passage at Ain Sukhna on 7th and 8th April 1995. An immature was found at Sharm el Sheikh in October 1995 with one wintering in 1997/98.

**Booted Eagle** *Hieraaetus pennatus* Up to a 1,000 birds have been seen passing Suez in the autumn with fewer birds recorded in the spring with 457 the highest count between March and May. Other sites have included Wadi Hagul, St. Paul's Monastery, Hurghada and Bir Beida, Quiser. Nile cruises have seen mainly individuals on many different parts of the Nile.

**Osprey** *Pandion haliaetus* A common breeding resident along the South Sinai and Red Sea islands with up to 200 pairs present. Nests contain many gar and pipe fish caught near the surface and used to add to the nest material. Hurghada is the best place to see this bird. Passage birds use the Nile Valley and small numbers winter with 17 counted in 1990, ranging from Lake Manzala to Lake Nasser. One wintered on Wadi el Ryan in 1997 and in Dakhla February 1998. Nile cruises regularly see this bird with a maximum of 21 birds along the river with Philae, Aswan a guaranteed site.

**Kestrel** *Falco tinnunculus* A common breeding resident with large passage right across the country. Many birds seen along the Nile with many observations seen around pyramids and temples where they often breed.

**Lesser Kestrel** *Falco naumanni* A passage migrant with a small casual breeding population in the north of the country. A single bird wintered in Wadi Feiran, Sinai in 1994/95. Small numbers seen on migration from Suez and small numbers also seen along the Nile. Like kestrels, passage birds can use the pyramids and temples to soar over and feed with 4-5 birds seen over Meydum and singles at Philae and the Valley of the Kings. A group of 6 were seen along the Zafarana-Hurghada road on 21st March 1995 with 3 on the Quiser-Qift road on 23rd March and 7 feeding around a mosque at Qus near Luxor on 15th April 1995.

**Red footed Falcon** *Falco vespertinus* This species is hard to see during migration due to its high-flying tactics and only small numbers have been observed with a maximum of 462 from Suez during a 2-month period in the autumn. Small numbers have been seen along the Nile with 15-20 coming to roost near Beni Suef during a birdwatching cruise, the highest count with singles the normal observation at Karnak, the Valley of the Kings, Hathor and Philae. An interesting observation came from Wadi el Natrun where two were seen catching frogs and toads on 16th May 1990.

**Merlin** *Falco columbarius* This is a winter visitor to mainly the northern half of the country with records from especially the Delta areas. A total of 17 were counted here in 1990 with passage birds seen twice at Wadi el Natrun. The northern part of Sinai has wintering birds and the only record from the Nile cruises came from El Amarna where one was seen flying with two Barbary falcon. A male was at Hurghada on 9th December 1997.

**Hobby** *Falco subbuteo* This is a passage migrant with a small breeding population on the north east border with Israel. In Israel there is a population of 3-5,000 pairs which all need the hooded crow's nest for breeding so why not a wider population in Egypt? Both El Arish and Rafa have had breeding birds. Passage is on a broad front with many records coming from the Nile cruises.

**Eleonora's Falcon** *Falco eleonorae* A high-flying migrant like the red footed falcon means that there are few records but birds are regularly seen from Suez in the autumn until the beginning of November. Single birds seen at Zaraniq on 16th September 1995 and on 2nd May 1996 high over the mountains above Taba, South Sinai. A single record of a possible light phase bird at Beni Hassan came from the Nile cruises.

**Sooty Falcon** *Falco concolor* A late summer breeding bird of mainly the eastern part of the country, relying on the autumn migration for feeding young. The maximum number breeding in one site is 150 pairs on Zabargat Island in the Red Sea. Up to 10 birds seen feeding on insects in Hurghada in 1990. Often seen on Nile cruises with a maximum of 9 birds seen with records from Meydum to Aswan. Several birds are found around Luxor with good sightings of birds coming to feed on roosting sand martins. One record from Cairo on 2nd February 1995 could be the first wintering bird in the Western Palearctic.

**Lanner Falcon** *Falco biarmicus* A resident breeding bird of the desert areas and along the edge of farmland. Commonly seen from the historic pyramids and temples where it also used to nest before the tourist came! Even mating has been observed at the Valley of the Kings and several birds have been watched while mobbing passing buzzards and eagles. Aswan seems the best bet for observing this species. Birds are also seen at Dakhla Oasis, Wadi el Natrun and passing through Suez on migration. The camel station near Abu Simbel had one bird, which had been captured using a snared pigeon in a trap called a sharak, being used as an exhibit (to photograph) by passing tourists.

**Saker Falcon** *Falco cherrug* An uncommon migrant and winter visitor to especially the Sinai and Eastern Desert. There are one or two records from Nile cruises with some birds seen very close, aiding identification. This species is still caught on migration for falconry usually by locals who then sell on the birds to Arabs from the Gulf states. Seven birds wintered in the north western Negev in 1996/97 and could have easily crossed the border into Egypt.

**Peregrine Falcon** *Falco Peregrinus* There are increasing numbers of birds reported for the country with several records from Nile cruises both seen along the river and around the sites visited including Valley of the Kings, Giza, El Amarna and Aswan. Most birds are passing through but some winter especially around Cairo where up to three birds wintered during 1994/95. One wintering birder set up a telescope in his high-rise hotel room to watch the birds raiding the local pigeon population. Recent transmitters have been placed on Siberian peregrines to track their migration, with birds so far being found along the Gulf of Aqaba and the Red Sea, especially on the Saudi side.

**Barbary Falcon** *Falco pelegrinoides* The records so far indicate a limited distribution for this bird with many breeding sites along the Nile Valley and scattered reports from around Hurghada down to the Sudanese border and Eastern Sinai. It has been seen in several localities on the Nile cruises with the most frequent records from the Valley of the Kings, El Amarna, and Aswan where a pair has been seen with individuals catching pigeons off the buildings. Old raven nests seem to be the preferred nesting sites.

**Chukar** *Alectoris chukar* A species restricted to the Sinai Peninsula where it does not seem common. Most records come from the north east corner from Wadi el Arish along to Rafa where adults and chicks have been seen Other past records come from wadis in the south of the peninsula. There is a strong population in Israel with 1,200 counted in the Western Negev in November 1993 and additional birds may well come over the border into Egypt.

**Barbary Partridge** *Alectoris barbara* Another restricted species with known sites close to the Libyan border along the Mediterranean. As few birdwatchers cover this area records are limited and thoughts are that the birds are declining, if not extinct.

**Sand Partridge** *Ammoperdix heyi* A common resident of the dry Eastern Desert and Sinai Peninsula but not always easy to see. Their habitat does not give them away but during early morning and late evening the birds can be heard calling to one another. Only one record exists from Nile cruises, from Beni Hassan. Most birders find this species along the Qena-Safaga road, the Quiser-Qus road and the Marsa Alam-Esna road where 4 adults were seen with 12 chicks on 1st May 1997. On Sinai, the St. Catherine area has been known to produce records.

**Quail** *Coturnix coturnix* This is a common migrant especially in the autumn when up to half a million birds can be caught along the Mediterranean coast for the markets around the country. It was found on Nile cruises and has been found along the Red Sea as well. Late birds may well breed in the spring but their numbers are limited.

**Water Rail** *Rallus aquaticus* A breeding resident of the northern area especially around the Delta Lakes with many migrants visiting the country, especially along the Nile. The species has been recorded from Nile cruises but its behaviour of skulking in vegetation has left most records as calling birds, especially in winter. Up to ten birds were calling at Lake Umm Beida on Wadi el Natrun in January 1997 while at first a marsh harrier flew over, followed by a hen harrier.

**Spotted Crake** *Porzana porzana* An unrecorded species which, like the water rail, often spends most of its life concealed in vegetation. There were no records from Nile cruises and passage birds have come from Sharm el Sheikh sewage ponds but the rest have been for sale as food in Port Said markets where up to 3 a day have been found. Far more have come from over the border at Eilat where up to 50 are counted during spring passage.

**Little Crake** *Porzana parva* There are many more records for this species than the spotted crake and there has been some evidence that the species has bred in the Delta Lakes. Passage birds are found down to Aswan and Nile cruises have spotted the bird but more often it has been placed as either this or Baillon's crake. Winter records have come from Crocodile Island at Luxor and the Al Alamein Hotel along the western Mediterranean coast and several have been for sale in the Port Said market. Lake Qarun has had the highest count of this bird with up to 6 in April. Gebel Asfar held 2 birds on 30th March 1995. One was at Sharm El Sheikh golf course on 20th February 1998.

**Baillon's Crake** *Porzana pusilla* A rare passage bird and a rare breeding bird of the Delta. A bird which is often confused with the little crake and some records are for either or. One of the few summer records is exactly that with a bird at Gezira Ghazal, Aswan on 7th July. The Nile cruise records are also of either bird.

**Corncrake** *Crex crex* The autumn population of this bird is heavily netted along the Mediterranean coast with up to 4,600 birds caught during 1991. There is little cover at this time of the year so the birds are more easily caught. This represents 20% of the Western Palearctic population which is declining due to modern agriculture. Due to winter rains, this area has plenty of cover to allow spring birds to return to Europe. Few records exist away from this coast but Nile cruises have seen this bird at Beni Hassan.

**Moorhen** *Gallinula chloropus* A very common bird found throughout the year with large numbers around villages where you would presume that the bird along with its eggs, would be commonly caught as food hence its English name. Commonly seen on Nile cruises often being dwarfed by its big cousin.

**Lesser Moorhen** *Gallinula angulata* The first record for the Western Palearctic was found at Abu Simbel on 6th May 1997. This African species adds yet

Hoopoe Lark at the Black Pyramid at Dahshur

another remarkable record from this site making it a special place for birdwatchers from around the world.

**Purple Gallinule** *Porphyrio porphyrio madagascariensis* A common breeding species found in most areas of the Nile Valley and Delta with other records from as far as Wadi el Rayan in the Western Desert and east across to Suez. Nile cruise boats act as wonderful platforms from which to look down on this amazingly big 'lily-trotter' which has enjoyed the spread of water hyacinth along the Nile with a total of 298 being seen on one trip. Other sites to see this bird are from felucca rides from Luxor and Aswan and from the corner of the hotel grounds on Crocodile Island.

**Coot** *Fulica atra* A rare breeding bird with a large influx of wintering birds. In the cold European winter of 1978/79 over 200,000 were counted on three Delta Lakes. Only 32,000 were counted in 1989/90. It is seen on many of the winter cruises along the Nile and some records even exist from the Red Sea.

**Common Crane** *Grus grus* A great spectacle of many of the Nile cruises is to witness the movement of this bird along the Nile. The distinctive call can be heard well before seeing the birds and they are much appreciated by birders escaping the cold and darkness of a northern winter. Up to 1,741 have been counted along the Nile with birds soaring over the Valley of the Kings and Philae. Away from here, there has been 1,725 counted in 4 hours at Ain Sukhna on 20th March 1992. 7,280 were counted in two hours, 30 kilometres west of Zafarana on 14th March 1996 - this represents the highest count in the country, so far, on one day. A staging post was once around Wadi el Natrun.

**Demoiselle Crane** *Anthropoides virgo* A bird which should be seen more often in Egypt with a large wintering population found in the Sudan. Several flocks were seen at Eilat with the last record there of a single bird on 20th/21st March 1995. The site which attracts annual birds is Cyprus with a maximum of 345 on 15th September 1995. This 'stopover' may well allow these birds to fly direct from Cyprus over Egypt and south to Sudan. One was at Sharm El Sheikh on 23/24th October 1997.

**Crowned Crane** *Balearica pavonina* There are some unconfirmed records of birds from Gebel Elba from the 1970s but if any were to be seen today the best bet would be at Abu Simbel!

**Little Bustard** *Tetrax tetrax* A former winter visitor with no recent records but there are still up to 100 wintering in Israel in the northern region.

**Houbara Bustard** *Chlamydotis undulata* A declining breeding bird due to heavy persecution from outside hunters using falcons. Only the border with Israel seems to hold any breeding birds. The last stronghold was on the Libyan border where annual culls have made the birds extinct. The future does not look bright unless someone takes hold of the situation and tries to save the remaining birds. It only needs a look across at Israel to see what can be done - up to 384 counted in Western Negev in November 1996. In contrast, a single bird was recorded at Ras El Naqb, north Sinai on 29th October 1994.

**Great Bustard** *Otis tarda* The first record for the country came from a bird shot at Salehia, 120 kilometres east of Cairo on 19th March 1992. The man did not know what the bird was so took it to have it stuffed where it was identified. A sad end for such a wonderful bird. It is a very rare bird also in Israel.

**Painted Snipe** *Rostratula benghalensis* A breeding bird of the Nile Delta as well as of Wadi el Natrun and of the Faiyum district. It is not always an easy bird to see and evenings give a better chance of seeing the bird move out into open areas to feed. Several sites are noted for this species including Gebel Asfar just outside

Cairo where a male was observed with chicks in 1990. Lake Qarun has been another area for observing them, especially around the drying fish ponds west of Shakshuk at Abuska Bay. Nile cruises have seen them near El Til and in 1997, two were seen on Crocodile Island, Luxor. A notable gathering was seen at Abu Grida near Lake Manzala on 13th May 1997 and at Abbassa Fish Farm in October 1997. The only passage bird recorded was a single bird at Sharm el Sheikh in September 1991.

**Oystercatcher** *Haematopus ostralegus* Small numbers winter with 127 birds in 1990 with few records away from the Mediterranean and Red Sea coasts. Passage birds included 151 at Lake Manzala and 56 at Suez during spring 1990. Normally twos and threes along the Red Sea coast and rare inland with singles seen by the Nile cruises south of El Fashn and the latest being at Dendera on 21st September 1993. Both Lake Qarun and Wadi el Natrun also have had birds.

**Black-winged Stilt** *Himantopus himantopus* One of the great show pieces from the Nile cruises. Nobody can imagine a bird having such legs with such a small body. A common passage bird but rare breeder with new sites found at 10th Ramadan City and 4 kilometres east of the Great Bitter Lakes where the total pairs in 1991 was 100. It is also a fairly recent breeding bird in Israel with 1-2,000 pairs present. A large winter concentration was found in Dakhla Oasis in February 1998 with 500+ birds present.

**Avocet** *Recurvirostra avosetta* A common winter visitor especially in the Delta with smaller numbers south on the Nile as far as Abu Simbel. Small numbers also seen at Lake Qarun and Wadi el Natrun. A total winter count of 21,814 birds was recorded in 1990 of which 20,000 were in the Delta. Birds were commonly seen on Nile cruises with migration seen at Aswan. New breeding evidence came from Lake Bardawil with 5 pairs in 1990 and 4 pairs in 1994 being the first breeding records for the country.

**Crab Plover** *Dromas ardeola* The major site for finding this species is down at Shelatine on the Red Sea where up to 180 birds have been seen. There were 38 there on 30th March 1997 with other sightings at Hurghada and a rare record of 2 on the beach of Lake Manzala on 30th April 1990. There are only 2 records from Sinai and one from Eilat but rarer ones from the Mediterranean Coast with 2 birds found even in 1997. Are these birds moving through the Suez Canal into the Mediterranean?

**Stone Curlew** *Burhinus oedicnemus* A breeding bird of the desert margins with records especially from the north of the country and along the Mediterranean coast. Nile cruises have found a number of birds and even a nest scrape at Beni Hassan. Other records have come from Meydum, El Til, Balliana and El Feshn. Birds have also been seen at Lake Qarun, Wadi el Natrun and Lake Manzala.

**Senegal Thick-knees** *Burhinus senegalensis* A common breeder of the Nile Valley and Delta with a new breeding feature being the nesting on flat roofs around Cairo where it is widespread. It is so funny to hear the birds over the noise of traffic or even glimpse them standing on the roofs as you drive by. It is commonly seen on Nile cruises with birds seen from many areas, with common views of them standing under trees and shrubs to keep cool in the shade.

**Egyptian Plover** *Pluvianus aegyptius* This bird was lost to the country with the making of the Aswan Dam which prevented any shingle from coming down the river where it nested. The waters behind the dam also stretched back into Sudan where the bird still can be found, leaving the only recent record from Wadi Halfa. Two birds were seen there in 1981 only 20 kilometres from the Egyptian border. There is an old record of a bird on Sinai in 1979.

**Cream-coloured Courser** *Cursorius cursor* A breeding bird of the desert areas especially along the coasts and where the desert meets the cultivated strips of the Nile. Wadi el Natrun and Lake Qarun are regular sites for this species with records also coming from the Cairo-Faiyum road. Birds have also been observed on the islands off Hurghada during the summer. The winter records are widely spread from the Mediterranean coast down to Lake Nasser where 8 birds were seen together at Garf Hessein. Passage has been recorded at many sites such as Lake Manzala where 449 were found on 12th May 1990, also at Port Said, Suez, Zaraniq and Gebel Elba where 21 were seen moving through the area on 30th March 1997. Nile cruises have seen the birds from Meydum down to El Amarna with a maximum of 11 birds seen flying in front of the boat.

**Collared Pratincole** *Glareola pratincola* A common breeding bird of the northern Delta Lakes with smaller colonies elsewhere. Over 2,000 pairs breed at Lake Burullus with another 500 pairs around Lake Manzala. Much smaller numbers are found breeding at Wadi el Natrun [10+ pairs] and Lake Qarun [15+ pairs]. It is a common passage bird with large numbers seen along the Nile, especially in the autumn with a maximum of 850 seen mainly in the Middle Nile with a day count of 720 birds at Minieh. Other records have come from Aswan, Abu Simbel, Suez and Hurgada rubbish tip and coast line. Small numbers have been seen at Sharm el Sheikh while up to 2,000 have been recorded at Eilat.

**Black-winged Pratincole** *Glareola nordmanni* An uncommon passage bird often overlooked in flocks of collared pratincoles. Recent views of this species at Hurghada rubbish tip had the two species well apart but records from the Nile suggest that birds were often feeding in similar areas. Records from Eilat estimate that for every 100 collared there are 1 or 2 black winged pratincoles. Other sites where this bird have been seen have come from Suez, Zaraniq, Sharm el Sheikh and Taba. The only records from Nile cruises have come from Aswan and near Luxor.

**Oriental Pratincole** *Glareola maldivarum* The first record for the country and third record for the Western Palearctic came on 2nd April 1993 at 10th Ramadan's sewage works which is 55 kilometres west of Cairo. The first record for Cyprus was on 24th April 1993 and could/will have been the same bird, given its rarity.

**Little ringed Plover** *Charadrius dubius* A rare breeding bird found around Wadi el Natrun with several pairs found in recent years. This bird is more of a passage migrant with records coming from throughout the country with high numbers at Hurghada rubbish tip [50+ on 1st April 1997], along the Nile [23 between Qena and Luxor] and 30 + at Lake Qarun. Winter records come from along the Nile Valley but especially Lake Nasser.

**Ringed Plover** *Charadrius hiaticula* A common passage bird and fairly common winter visitor with up to 2,000 wintering in various parts of the country but only 970 in 1990. It is commonly seen along the Nile depending on the water levels, and large numbers pass through Suez Bay in especially April when counts have shown up to 1,300 birds on one day.

**Kittlitz's Plover** *charadrius pecuarius* A rare breeding bird which most birders have on their list to see in Egypt. Around 200 pairs breed on Wadi el Natrun with 40+ wintering there, making this the easiest place to find this species. Other records come from the Delta Lakes with Lake Manzala and Lake Qarun outstanding, but Abu Simbel held 36 birds in 1997 with breeding confirmed from here as well. Pairs were also located north of Abu Simbel in 1997 and with such a large surface area to cover, several pairs may breed here away from the normal sites. Nile cruises have come across this bird especially at Kom Ombo and Aswan. Recent records in Israel are due to a supposed increase in breeding birds in Egypt!

**Kentish Plover** *Charadrius alexandrinus* An estimated total of 5,000 pairs may breed in Egypt with 3,000 pairs based around the Delta Lakes and Lake Bardawil. Wadi el Natrun holds up to 200 pairs and smaller numbers are found at Lake Qarun, Wadi el Ryan and the Bitter Lakes. Winter can hold up to 10,390 birds with the bulk of these again around the Delta Lakes with 4,323 at Lake Manzala. A common passage bird with other records coming from around the country with a special mention for the Nile Valley, Abu Simbel, and Hurghada.

**Lesser Sandplover** *Charadrius mongolus* A rare passage migrant with many new records as birdwatching increases around the country. There were 3 at Suez on 20th March 1990, one at Hurghada on 17th March 1992 with another at Lake Qarun on 7th April 1992 and again on 5th February 1993; another at Suez on 13th December 1996 and 15 at Shelatine on 1st April 1997 where a maximum of 20 have been counted here before. There have been as many as 5 records for Sinai but only one for Eilat.

**Greater Sandplover** *Charadrius leschenaultii* A common wintering bird with up to 1,000 birds estimated. Only 128 were counted in 1990 with 75 of these in Suez Bay and 24 at Lake Burullus. Small numbers are found around Lake Qarun and Wadi el Natrun with passage birds observed along the Red Sea with 137 at Hurghada on 6th September 1996. A rare bird along the Nile with cruises seeing this bird only at Kom Ombo and Beni Hassan. Sinai has held flocks of up 130 birds, especially at Nabq with smaller numbers wintering.

**Caspian Plover** *Charadrius asiaticus* There were 6 at Hurghada on 15th April 1990, one at Sharm el Sheikh on 13th March 1992 and one at Hurghada on 7th May 1996, making this a rare passage bird for the country. This is mainly a spring migrant at Eilat where some years up to 400 birds can be found.

**Dotterel** *Charadrius morinellus* A rare passage bird with former records suggesting wintering in the north east Sinai. There can be around 550 birds wintering in the north west Negev next door in Israel in recent years, so future finds are very likely in the Sinai. Other records have come from Sharm el Sheikh, Baltim and Wadi el Natrun with a single record from Gebel Asfar on 15th November 1991.

**Golden Plover** *Pluvialis apricaria* A scarce winter visitor with few recent records. Birds have been seen on Lake Qarun, Wadi el Natrun and at Zaraniq. Nile cruises have had only one record with 6 flying through Aswan on 17th January 1993. Up to 300 can winter in north west Negev with some birds thought to cross the border. The earliest autumn dates have been 6th August and a late spring record is 6th May.

**Pacific Golden Plover** *Pluvialis fulva* A rare and mainly autumn-occurring bird with singles from Lake Bardawil on 27th August 1981, 18th and 21st August 1990 and a full summer bird on 25th September 1990 at Lake Qarun. There have been 24 records in Israel with 19 occurring in the autumn.

**Grey Plover** *Pluvialis squatarola* This bird is mainly a winter visitor with up to 2,000 wintering, especially along the Delta Lakes with 1,500 found at El Malaha in 1990 along with 250 on Lake Manzala and 180 on Suez Bay. Passage birds have been seen along the Red Sea with gatherings again at Suez Bay of up to 360 but 2,690 counted near Umm Khalef, Lake Manzala is a record number for the country. Inland birds include Bahariyya Oasis, Lake Qarun, Wadi el Natrun and several records from Nile Cruises with birds seen as far down as Aswan.

**Three banded Plover** *Charadrius tricollaris* First for the country and the Western Palearctic was in the period 5th-26th March 1993 at the sewage works at Gebel Asfar just outside Cairo. Another bird was found at Aswan on 14th December 1997.

**Spur winged Plover** *Hoplopterus spinosus* One of the many highlights while birding in Egypt is this species which is very common around agricultural land. The boats on the Nile very rarely bother it as they glide by while they go about their daily chores. Winter birds can concentrate at such sites as Gebel Asfar, Wadi el

Natrun in the spring months and Crocodile Island near Luxor. New areas recently colonised are along the northern Sinai coast between El Arish and Rafa with small numbers wintering in southern Sinai.

**Black headed Plover** *Hoplopterus tectus* Only the second record for the Western Palearctic was found at Eilat on 11th April 1995.

**Red Wattled Plover** *Hoplopterus indicus* One was seen at Eilat from 28th December-17th February 1992. It is a breeding bird from the Indian subcontinent and has been seen also in Turkey.

**Sociable Plover** *Chettusia gregaria* Formerly a bird often found in lapwing flocks with records from many northern sites but records have decreased in recent years even though 19 were found wintering in the north western Negev during 1996/97. Nile cruises have had a single bird only, seen with lapwing at Lake Qarun. The last record was of a single bird found at the Dakhla Oasis, Western Desert on 22nd March 1993.

**White tailed Plover** *Chettusia leucura* A rare wintering bird especially of the Upper Nile with 62 counted between Nag Hamadi and Aswan. A flock of 30 birds was exceptional in this count. It is commonly seen from Nile cruises from where this high count came. Crocodile Island, Luxor is a regular site to see this species with up to 7 birds using the southern end of the island. An early autumn record of two was found at Zaraniq on 19th September 1994. Other sites where this bird can pass through include 10th Ramadan City sewage farm with one there on 27th September 1995 and Lake Manzala where one was there on the same date with a late spring record from here on 13th May 1997 at Abu Grida. The Red Sea coast has had several birds at the Hurghada rubbish tip while 2 more were found 10 kilometres south of Safaga on 10th October 1995. Up to 8 birds have been found at Sharm el Sheikh where again it is a passage bird. There were 4 at Dakhla and 2 at Farafra in February 1998.

**Lapwing** *Vanellus vanellus* A wintering bird of mainly the Delta area with some records from Faiyum, and the Nile counts have been up to 2,348 birds, with over 900 of these coming from Lake Qarun in 1990. Nile cruises have spotted very small numbers along the banks with an exceptional southern record of birds flying south at Aswan. Few records come from the Red Sea or South Sinai.

**Great Knot** *Calidris tenuirostris* A single bird found at Eilat on 24th October 1985.

**Knot** *Calidris canutus* A rare spring and autumn passage bird with most records coming from North Sinai with no more than ten birds found in recent times at Zaraniq.

**Sanderling** *Calidris alba* A fairly common migrant on mainly the North Sinai coast where up to 5,000 birds have been counted in the autumn. Smaller numbers winter, with up to 100 counted in the same area. There are one or two inland records with birds found at Wadi el Natrun and Lake Qarun, with Nile cruises finding a single bird in the spring near Beni Suef on 23rd May 1992.

**Little Stint** *Calidris minuta* A common passage and winter visitor with an estimated population of 100,000 birds wintering. Only 22, 608 were counted in 1990 with most of these found in the Delta Lakes. It is a common bird along the Nile with Lake Qarun especially popular.

**Temminck's Stint** *Calidris temminckii* A rarer passage and winter visitor with numbers rarely counted in double figures. They prefer fresher water than their cousins and could be found feeding in a stream running into Lake Qarun rather than along the lake shore in 1997. The drying fish ponds in Abuska Bay, Lake Qarun are a popular site for them where they can be seen feeding with little stints. A total of 17 were found wintering in 1990. Nile cruises have usually found this bird at Kom Ombo with a total of 3 found in winter on 14th December 1990, close to Beni Hassan. In North Sinai, Lake Bardawil has an old record of 257 counted there.

**Long-toed Stint** *Calidris subminuta* The first record for the area was photographed at Eilat on 25th/26th August 1991.

**Pectoral Sandpiper** *Calidris melanotos* A single bird was found at Lake el Zugm which is part of the Wadi el Natrun depression. It was feeding with other waders and was seen only on that day of 15th May 1990. There have been only 4 records in Israel with 3 of these at Eilat.

**Curlew Sandpiper** *Calidris ferruginea* An uncommon migrant both in spring and autumn with highest counts coming from North Sinai with up to 600 birds in the autumn and 1,255 at Lake Manzala in spring. It was found on Nile cruises in small numbers with the best sites being Lake Qarun and Kom Ombo.

**Dunlin** *Calidris alpina* A very common passage and wintering bird of the Delta Lakes and Nile Valley. Like little stint there is an estimated 100,000 birds present in winter with only 24,662 counted in 1990 with 11,275 coming from Lake Manzala. It was commonly found on Nile cruises with Lake Qarun especially good for this species.

**Broad-billed Sandpiper** *Limicola falcinellus* With recent work carried out in the Delta Lakes, larger numbers of this species have been found moving through the country between 5th and 21st May. Up to 80 birds were counted with many of these staying only for a day, making the total more like several hundred. Most of these birds were recorded in Lake Manzala or Suez Bay. Autumn records also come from Zaraniq with one there on 4th September 1995 followed by four there on 16th September and one on 19th September 1996. Red Sea records are scarce with three on 3rd May 1997 near Hurghada, a rare record. Few records come from South Sinai with the highest being 7 at Nabq on 2nd May 1976.

**Ruff** *Philomachus pugnax* A commonly-found passage and winter visitor with many records coming from Nile cruises with winter records from Crocodile Island, Luxor. In a total winter count of up to 800 birds most of these were from the Delta Lakes with 292 at Lake Manzala and 171 at Lake Qarun counted in 1990. Passage showed up to 3,000 birds at Lake Manzala on 25th March and 8,409 on 9th April 1990. No counts were carried out on agricultural land where this bird often feeds.

**Buff breasted Sandpiper** *Tryngites subruficollis* There is a very old record from Quiser, Red Sea, in 1928.

**Jack Snipe** *Lymnocryptes minimus* An uncommon wintering bird of the northern half of Egypt with records coming from the Delta Lakes, Lake Qarun and the Nile Valley (Crocodile Island). It is not a bird which can be seen well due to its crouching behaviour, so a single record comes from the Nile cruises only from a walking trip at Lake Qarun. Up to 17 were counted in the winter of 1990 with Wadi el Natrun holding as many as 7 birds.

**Common Snipe** *Gallinago gallinago* A common winter visitor which is very hard to count unless flushed. Several thousand probably use these extensive wetlands. A total of 552 were counted in 1990 with Wadi el Natrun holding 121 birds. It seems to be a very early migrant in the spring with most birds leaving in March.

**Great Snipe** *Gallinago media* A rare passage bird with small numbers wintering with one at Lake Idku on 17th January 1990. There was a record of a single bird at Wadi el Natrun on 8th March 1992 and another from Lake Manzala on 15th

March 1997. A possible sighting of a single bird came from Nile cruises at Kom Ombo on 23rd March 1986. A single bird was at Abbassa on 6th October 1997.

**Pintail Snipe** *Gallinago stenura* A single bird was ringed near Eilat on 1st November and stayed around until 7th November 1984.

**Woodcock** *Scolopax rusticola* A rare winter visitor to the north coast of the Mediterranean with few inland records as far as Giza and in South Sinai from St. Catherine's down to Sharm el Sheikh.

**Black tailed Godwit** *Limosa limosa* A regular winter visitor to the Delta Lakes with up to 1,335 counted in 1990 and 935 found on Lake Manzala. In the spring a small passage takes place with up to 300 birds observed with smaller numbers at Lake Qarun where 48 were counted on 30th March 1997. Nile cruises have found the bird at many sites with migration watched at Aswan, as birds headed south.

**Bar tailed Godwit** *limosa lapponica* A total of only 43 were found wintering in 1990 whereas around 127 were seen during the spring at Lake Manzala. Suez is the main site for seeing this bird on the Red Sea. Inland records have shown birds at Wadi el Natrun, Lake Qarun and on the Nile.

**Whimbrel** *Numenius phaeopus* The North Sinai coast has good numbers in the autumn with smaller numbers migrating through the country in the spring mainly along the Red Sea coast but inland records have come from Lake Qarun with a single bird, with curlew there on 30th March 1997. Nile cruises have seen few birds but 4 flying south at El Minieh on 21st August 1993 were exceptional. Possible wintering birds were found at Shora el Mantaqa, South Sinai on 4th January 1995.

**Slender billed Curlew** *Numenius tenuirostris* With a declining population thought to be on the edge of extinction, only one recent record from 1982 at Zaraniq seems worth mentioning. The small wintering population which used to come to the North African coast stretching from here to Morocco seems to be down to one or two individuals and even these cannot be found in some years.

**Curlew** *Numenius arquata* A small wintering population of only 500+ exists in a few localities and even these seem to take the brunt of the local hunters. In 1990 only 381 were found mainly at Wadi el Natrun with 145 birds and El Malaha with 104. Inland records come from Lake Qarun and the Nile Valley where several records come from cruises including a flock seen from Philae on 30th August 1993 of 30+ birds. Some birds are known to summer along the Red Sea coast.

**Spotted Redshank** *Tringa erythropus* This is a fairly common passage bird with a small wintering population especially around Lake Qarun where up to 46 birds were seen on 11th January 1997 around the fish pools at Abuska Bay. Spring migration held 196 birds on Lake Manzala on 3rd April 1990 with other records coming from the Nile cruises, Suez and Wadi el Natrun.

**Redshank** *Tringa totanus* Of an estimated winter population of 10,000 a very close count of 9,869 was counted in 1990 with over 7,000 of these coming from Lake Burullus and Lake Manzala. The birds were commonly found at Lake Qarun and on Nile cruises with records from many parts of the Nile. Birds are also found along the Red Sea coast with birds at Hurghada rubbish tip in 1997.

**Marsh Sandpiper** *Tringa stagnatilis* This is a scarce winter visitor with a larger population moving through on passage. An estimated 400 birds winter, with 98 being found in 1990 mainly in the Delta Lakes area with smaller numbers along the Nile and Lake Qarun. The bird was regularly found on Nile cruises and records come from as far as Abu Simbel where up to 10 birds were found during January 1997. The highest spring count was of 161 at Lake Manzala while autumn counts of 209 came from Lake Bardawil. Several records come from the Red Sea with birds

Crocodile Island-Squacco Heron,Whiskered Tern, Catlle Egret

Squacco Heron

Purple Gallinule/the big Nile

Felucca

Felucca

Felucca

Gebel Selin

Skink

Squacco Heron

Little Egrets

Glossy Ibis

Marsh Harrier

Black Stork

El Amarna

Ruddy Shelduck

White tailed Plover

Ruff

Spotted Redshank

Spur winged Plover

Water Buffalo on the Nile

Turtle Dove

Blue cheeked
Bee-eater

Pied Kingfisher

Strong House Sparrows!

Rufous Bushchat

Upstream from the Old Cataract Hotel

Kitchener's Island Arboretum

Agha khan Mausoleum

Philae

Golden shower flowers

Downstream from the Old Cataract

White Stork

Little Egret

Black Kite

Cattle Egret

Nile Acacia flowers

found at the Hurghada rubbish tip in April 1997 and at Suez.

**Greenshank** *Tringa nebularia* Surprisingly few birds winter in Egypt, with 500 estimated but only 144 counted in 1990. It is a very common bird on passage with many records coming from the Nile. A total of 543 were counted on one trip in the autumn with one stretch coming up with 110 birds. Their distinctive call makes it a bird not to miss. The bird is common on the Sinai with most wintering birds coming from the south where the mangroves offer easy feeding.

**Green Sandpiper** *Tringa ochropus* A very common passage bird with a wintering population of only 500+ with only 84 counted in 1990. The bird's feeding habit of moving well away from main lakes and concentrating on small canals such as the Faiyum area makes accurate counting impossible. It was very common at Abuska Bay, Lake Qarun in the spring 1997 with the highest autumn counts coming from Lake Bardawil where up to 987 birds have been counted moving through. A very common bird on Nile cruises seen all along the river.

**Wood Sandpiper** *Tringa glareola* A fairly common passage bird with increasing numbers of birds wintering. Groups of 30-40 birds were seen at Abuska Bay, Lake Qarun, Wadi el Natrun, Crocodile Island, Luxor and Gebel Asfar in January 1997. Very common on passage with birds found throughout the country and Nile cruises often finding this bird.

**Terex Sandpiper** *Xenus cinereus* One of many species which are increasing due to many more birders covering the ground. In 1990 there were 11 sightings with further ones coming from Suez on 12th April 1993; one also there on 8th April 1995, one on 14th September 1995 at Zaraniq, a single bird at Suez on 16th March, five at Zaraniq on 29th August 1996, one at Hurghada on 9th April 1997 and one at Ras Mohammed, South Sinai on 17th May 1997. The largest flocks have included 25 birds off Lake Bardawil and 15 at Shalatein, Red Sea. Only two records have come from Nile cruises at Aswan and Qena.

**Common Sandpiper** *Actitis hypoleucus* A common migrant with a small number wintering with an estimated 500 birds but only 77 were counted in 1990 It has been a regular on cruises on the Nile where most of the birds were found in the winter counts. Up to 60 birds have been counted along the Nile during passage from the cruises and other birds have come from many sites including the Delta Lakes and Lake Bardawil where up to 200 can pass through during August and September.

**Turnstone** *Arenaria interpres* An uncommon passage and winter visitor to beaches around the Red Sea and Mediterranean with a few inland records. Several records have come from Nile cruises with records from Lake Qarun and a single one from the Nile itself near Tima on 23rd September 1993, with an older record from Aswan in 1984. Up to 4 birds were feeding on Gifton Island, Hurghada on 29th January 1997. Passage can be seen from Suez and Lake Bardawil most years.

**Red necked Phalarope** *Phalaropus lobatus* There are increasing records in recent years of this species with Lake Bardawil standing out as a site for this species. Up to 25 birds were found on 7th September and 8th October 1995 here with a spring count of 12 birds on 9th May 1997 and the earliest passage record of 2 birds here on 20th and 21st March 1997. Work carried out west of the Suez in 1990 brought a total of 20 birds mainly from Lake Manzala with a total of 7 there on 24th March. A winter record from Lake Qarun on 21st January 1990 was also unique. There is a single record of 2 birds at Hurghada on 6th May 1991 and another inland record of several birds at Wadi el Natrun.

**Grey Phalarope** *Phalaropus fulicarius* A rare visitor to mainly the Mediterranean coast with several records from the Lake Bardawil area. A single bird was found here at Zaraniq on 20th September 1995. Several records come from Eilat so the Red Sea coasts are well worth watching for this species, especially at Ras Mohammed.

**Skua sp** *Stercorarius* Formerly a rare bird in Egypt but now a regular at Suez if you watch out for other birds besides raptors! There is a regular migration route of skuas moving up the coast with a total of 153 counted in 1990. The movement takes them inland to cross into the Mediterranean, passing on a broad front and this passage needs further examination in an attempt to put some real figures to the number of skuas passing through Egypt. Like the newly-found British passage along the Solway, numbers can vary so much depending on weather patterns which help only if birds are pushed down to sea level where they can be counted.

**Pomarine Skua** *Stercorarius pomarinus* Small numbers are regularly seen off Suez with up to 9 in 1990. Winter records come from Lake Nasser with a single on 24th December 1989 near Kalabsha Temple and Port Said with 2 birds on 28th January 1995. Port Said is the other site used by this species with several spring and autumn records, with Lake Bardawil also finding birds.

**Arctic Skua** *Stercorarius parasiticus* Increasing numbers of birds are now counted at Suez with 112 passing through between 27th March and 13th May 1990. Other records come from the Mediterranean where a build-up of autumn migrants takes place around Port Said with up to 34 counted. Larger numbers are counted at Eilat with up to 1,300 moving north in spring. Winter records also come from the Mediterranean especially at Port Said where 7 were recorded here in 1990. A bird ringed in Shetland, UK was picked up in Sudan on the Nile showing yet more inland movement.

**Long tailed Skua** *Stercorarius longicaudus* Single birds are the majority of records of this rare species to the area with one seen off the Suez on 5th May 1990 with another at Ras El Hikma and Zaraniq on 2nd and 6th October 1994 respectively. Port Said held two birds on the 20th February 1995 and there is an old record of 4 birds at Lake Bardawil in 1981. Eilat can have between 30 and 90 birds during the spring passage.

**Great Skua** *Stercorarius skua* Very few records exist of this species with confusion with the south polar skua. A single bird was seen at Lake Bardawil as late as 1972 with other records of skuas from Port Said may be of this species. Only 4 records exist for Israel with one of these coming from a ringed bird from Iceland back in 1966.

**South Polar Skua** *Stercorarius maccormicki* Following the confusion between this and the great skua, only one confirmed record exists from Gifton Island off Hurghada from a bird tour run from Eilat! The bird was seen on 10th June 1991. Only two records are confirmed from Eilat with a single bird there from 3rd to 6th June 1983 with another here on 28th June 1992.

**Sooty Gull** *Larus hemprichii* A rare breeding bird of the Red Sea islands with as few as 50 breeding pairs of Hurgada. No breeding birds were found in the summer of 1994 and it was thought that the survey had started too early for this species or that the white eyed gulls had taken over their former breeding grounds. A number of these gulls were watched trying to predate the white eyed gull chicks. This species is commonly seen at Hurghada and on the boat trips to the nearby islands. Other birds breed farther south from Wadi Gemal down to the Gezira Siyal group along the Red Sea. Sinai has several records from Ras Mohammed while 15 were north of Sharm el Sheikh on 15th June 1990 at Ras Nusrani and three more at Nabq on the same day. Six were counted at Nabq on 5th April 1994. There are only a handful of records from Eilat.

**White eyed Gull** *Larus leucophthalmus* A population of around 1,500-2,000 pairs are thought to breed along the Red Sea coast. A survey in June 1994 found few pairs breeding and no information indicated the number of pairs present. This bird is far commoner than the sooty gull and is easily viewed even from the shore at Hurghada where 6,500 were counted at the rubbish tip here on 7th May 1996. This represents the largest concentration in the country (or the world?). A single bird was seen at Port Said and is a rare record for the Mediterranean on 29th January 1995.

**Great Black headed Gull** *Larus ichthyaetus* A widespread wintering species with a total of 35 counted in the northern region in 1990 with records inland at Wadi el Rayan with a maximum of 19 on 22nd January 1990 and 17 there on 28th January 1995. The best count of 20 birds was seen on 30th December at Lake Qarun with only 8 present in January 1997 rising to 10 on 24th February with a record 22 on Lake Nasser. Records from the Nile cruises include a group of 4 at Edfu on 29th March 1989. Passage has had 18 adults pass Suez between 4th and 21st March 1990 and 10 on 7th February 1997.

**Mediterranean Gull** *Larus melanocephalus* An increasing winter visitor along the Mediterranean coast with 91 birds at Port Said in January 1990 and 80+ at Ras el Bar on 19th February 1997 with more than 50 adults and many first year birds mixed in with black headed gulls and sandwich terns. Other records include 7 birds found at Port Said on 20th September 1994 but 120 there on 20th February 1995. The Red Sea coast is also known to hold birds in small numbers with records from Hurghada in 1994 and inland records from Lake Qarun and the Nile Valley.

**Little Gull** *Larus minutus* Once thought to be a scarce wintering bird but a count in 1990 produced 52,769 birds of which 47,316 were found on Lake Manzala. This represents the highest total of birds counted in the world at a wintering ground which used to be thought of for whiskered terns, not this species. The birds may well have come from the sea close by and/or moved down from the Black Sea wintering population. At least 5,000 birds were at sea off Port Said on 23rd March 1990. Very small numbers have been seen elsewhere with records from Lake Qarun, the Nile Valley, the Red Sea coast and Sinai.

**Sabine's Gull** *Larus sabini* A very rare migrant with only two records from Zaraniq up to 1980 but a single bird seen 77 kilometres west of Alexandria on 2nd March 1995 was the third record for the country. There are two recent records for Eilat with 1st summer birds appearing from 3rd July to 2nd August 1989 and another on 17th June 1992.

**Black headed Gull** *Larus ridibundus* A very common wintering bird with 112,253 found in 1990, with 64,543 of these found on Lake Manzala feeding with little gulls and whiskered terns. Another count comes from the Nile cruises where 10,000 were counted in March between Cairo and Aswan. Migration also occurs along the Red Sea where over 400 birds were moving north by 22nd February 1997 with several oiled birds present. Sinai also has a passage but it does not match any of the other areas.

**Brown headed Gull** *Larus brunnicephalus* A single 1st summer bird was found at Eilat on 12th May 1985.

**Grey headed Gull** *Larus cirrocephalus* Three records all from Eilat in 1989 with a single first summer, a single second summer and a single second winter bird.

**Slender billed Gull** *Larus genei* A breeding bird of the Delta Lakes with 5,688 nests counted on El Malaha in 1990 and 11,251 birds counted in the same winter with most of these from the Delta Lakes and a good count of 1,332 from Lake

Qarun in March 1990. Passage has been observed at Suez with small numbers present into April at Hurghada. Nile cruises have seen the birds along the Nile especially the Upper Nile and on the lake around Philae.

**Audouin's Gull** *Larus audouinii* This bird's increasing breeding population in the Mediterranean has resulted in many more records in recent years, with 2 found at Montaza Harbour, Alexandria on 23rd January 1993. A first winter bird at Marsa Matruh on 19th March 1994, a single adult found at Port Said on 20th September 1994, up to five at Zaraniq from the 7th September to 8th October 1995, a maximum of 13 on 14th September at Zaraniq and another 7 up to 3rd October 1996, making Zaraniq the most consistent site for finding these birds in the country. There are up to 6 records from Eilat in recent years.

**Common Gull** *Larus canus* This a rare winter visitor with few records of late. There was a single record from Abuska Bay, Lake Qarun on 21st January 1990 with another at Port Said on 11th March 1990. Another bird was found on Lake Qarun during March 1993 and another record there in January 1997.

**Lesser black backed Gull** *Larus fuscus* A common passage bird with several thousand counted from Zaraniq in the autumn. Birds then spread right across the country with smaller numbers wintering; 119 were counted around the wetlands of Egypt in 1990 with over 2,000 migrating past Suez in the spring of 1990. It is commonly seen on Nile cruises with records as far south as Aswan.

**Siberian/Heuglin's Gull** *Larus heuglini* A recently-created species with its background divided between several of the big gulls, especially lesser black backed gull. It is fairly common in Israel with scattered records for Sinai. A single bird was seen in Alexandria harbour on 18th January 1990.

**Yellow legged Gull** *Larus cachinnans* The commonest large gull wintering mainly around the Delta Lakes with few records inland. Up to 2,290 were found in 1990 with 1,440 of these on Lake Manzala. Port Said is a site from where they can be easily observed, with Suez offering birds in winter and on passage. A small population breeds along the Mediterranean coast with their eggs much in demand for eating by locals. Inland records have been at Lake Qarun and several records on Nile cruises as far south as Aswan.

**Armenian Gull** *Larus armenicus* A regular wintering gull of the eastern Mediterranean coast and along the Red Sea. Highest counts come from Lake Manzala where there were 253 in 1990 and 520 on 15th March 1997. Several have been seen at the Great Bitter Lakes where up to 22 were counted in 1990 and up to 10 have been counted at Suez and Ain Sukhna. They have been recorded as far south as Hurghada. Sinai records chiefly come from Lake Bardawil in the north.

**Herring Gull** *Larus argentatus* An extremely rare bird with only one record for Israel with no records yet accepted in Egypt but 3 birds were seen along the Red Sea in May 1991.

**Great black backed Gull** *Larus marinus* No past records for this bird have been accepted for the country but an adult was at Lake Nasser on 22nd January 1989 and 49 birds were counted mainly at 3 sites in 1994 with Lake Qarun, Wadi el Ryan and Nasser again taking the prizes. The 5 adult birds seen at Hurghada on 12th April 1995 now do not seem so strange.

**Kittiwake** *Rissa tridactyla* A rare bird for Egypt with scattered records mainly from the Red Sea where numbers move through the Gulf of Aqaba to Eilat. Most records are of immature birds with a first summer bird at Ain Sukhna on 5th May 1990 and at Hurghada on 9th April 1997. One or two records come from the Mediterranean coast, especially around Port Said.

**Gull billed Tern** *Gelochelidon nilotica* A fairly common passage migrant with small numbers wintering along the Nile and on Lake Nasser with birds seen in 1997 at Crocodile Island, Luxor and Abu Simbel. This was frequently seen on Nile cruises with several records from Philae and Aswan. Up to 20 birds were feeding at Philae on 6th April 1997.

**Caspian Tern** *Sterna caspia* A fairly common breeding bird of the Red Sea islands with up to 350 pairs breeding where birds tend to remain all the year round. Ringing returns have shown a large population migrate through Egypt from Scandinavian countries with up to 85 birds moving past Zaraniq in the autumn. Inland records have come from Lake Qarun and the Nile Valley where cruises have contributed to the records. The easiest place to see this species is at Hurghada and Sharm el Sheikh where there is another small breeding population on Tiran Island.

**Crested or Swift Tern** *Sterna bergii* A regular species found off the Red Sea coast with small numbers present throughout the year. There is no proof of breeding yet but birds are found along with other nesting terns in suitable habitat off the Hurghada Islands and Tiran Island. Movement along the Red Sea has been recorded at Suez. Groups of up to 50 birds have been seen between Ras Mohammed and Attur during April and October with only 10 records from Eilat.

**Lesser Crested Tern** *Sterna bengalensis* This is a summer visitor to the Red Sea area with up to 2,000 pairs breeding with a further 100 pairs on Tiran Island off Ras Mohammed. There were three birds in Suez Bay on 2nd January 1990 which was the first winter record for this species. There is a small movement along the Suez Canal with the only inland records coming from the Nile cruises with 2 birds seen well near Beni Hassan on 28th September 1986. An even older record comes from Khartoum, Sudan on June 26th 1982 when the bird was captured.

**Sandwich Tern** *Sterna sandvicensis* A fairly common migrant with winter populations as high as 1,596 in 1990, with up to 574 birds on Lake Burullus and 801 on El Malaha. Migrants were seen at Suez in April 1990 with the only inland record coming from the Nile cruises with a single bird seen well, south of Nag Hammadi on 2nd October 1985.

**Roseate Tern** *Sterna dougallii* Of the records known about in Egyptian waters none have been accepted as yet but they are worth mentioning in case someone else has records which have not yet been submitted. Five birds were seen at Suez in 1980 and 2 Nile records from cruises came from Beni Hassan on 16th March 1985 and 2 possible birds near Sohag on 30th September 1986 with the caption 'almost certainly this species'.

**Common Tern** *Sterna hirundo* A new breeding bird for Egypt with 2 eggs and 11 screaming birds found at El Manaha on 26th May 1990. It is a common passage bird with up to 12,000 counted at Zaraniq but an amazing 20,147 were counted between 19th March and 17th June 1990 at Suez. Inland records have come from many parts of the Nile with records coming from the cruises. Summering birds have also been found around Hurghada.

**Arctic Tern** *Sterna paradisaea* Two birds were seen off Tawila Island, northern Red Sea on 16th July 1996. This may well be the first record for Egypt even though Eilat has had a number of birds since 1982 when their first confirmed record was found. A single bird was captured at Khartoum, Sudan well inland on 16th May 1982.

**White cheeked Tern** *Sterna repressa* A breeding bird of the Red Sea with up to 150 pairs using the islands off Hurghada, with birds turning up in the first weeks of April and remaining until late October with last records in early November. Movement can be seen at Suez with birds seen as far as Ismailiya with one record from the

Mediterranean at Jaffa Port, Tel Aviv on 10th January 1986.

**Bridled Tern** *Sterna anaethetus* A summer migrant to the Red Sea with up to 300 pairs thought to breed especially around the Hurghada Islands. Many birds were present in June 1994 with up to 40 pairs present on some islands with eggs yet to be laid. They may winter in the southern Red Sea with records from Gezira, Rawabel el Kabir and Shalatein. Up to 30 have been seen off South Sinai with more records at Eilat in recent years

**Sooty Tern** *Sterna fuscata* Up to 19 birds all in adult plumage have been seen off Eilat, all in high summer being late July and early August, with odd birds seen from pelagics off the Sinai, Red Sea in the 1990s with birds also from the Sudanese coast.

**Common/Brown Noddy** *Anous stolidus* A single record comes from the Nile at Edfu on 23rd March 1986 with the bird seen by a Nile cruise. The account sounds as if several people saw the bird as it perched on a sandbank as the boat drifted by. This species is supposed to breed off the Sudanese coast.

**Little Tern** *Sterna albifrons* A substantial number of the Mediterranean's breeding population are found in Egypt with recent work pushing the numbers of birds up to 3,800 pairs, with up to 1,500 pairs found on Lake Manzala alone. Another large count is found around Lake Bardawil with up to 1,200 pairs in 1990. Large movements of birds were recorded from Suez with 9,697 birds counted between 20th March and 29th May 1990. Inland birds were found breeding at Lake Qarun and Wadi el Natrun with up to 80 pairs and 55 birds seen at Abu Simbel on 10th July 1990. Nile cruises have seen several birds along the Nile and birds have also been recorded from Hurghada. Given recent figures, Egypt now has at least 18% of the Western Palearctic population and up to 40% of the Mediterranean breeding population.

**Saunder's Little Tern** *Sterna saundersi* There are increasing numbers of records due to bird tours out from Eilat to look for seabirds with 5 in the northern Gulf of Suez on 18th September 1991 followed by 2 seen from land at Ras Mohammed on 10th April 1992. A single bird was seen on Tawila Island, Red Sea on 30th August 1995 and 3 possible sightings off the northern Red Sea between 13th and 18th July 1996. There are no records from Eilat since July 1989. This species may breed in the southern waters of the Egyptian Red Sea.

**Whiskered Tern** *Chlidonias hybridus* A very common winter visitor with an amazing 44,484 birds counted in 1990 with 38,741 counted on Lake Manzala. This represents the main population of the Western Palearctic east of Spain where the birds migrate down West Africa. A common bird seen on Nile cruises with several records coming from Kom Ombo and Philae. Small numbers are found along the Red Sea with Hurghada rubbish tip standing out as well as Suez with counts of up to 1,000 birds from Zaraniq in the autumn.

**Black Tern** *Chlidonias niger* This is the rarest of the three 'marsh terns' with the largest counts coming from Lake Bardawil with up to 240 having been counted. Wintering birds were found on Lake Nasser where up to 118 were found in January 1980. Single wintering birds have been found on Lake Qarun and on the Nile. Small numbers have been found by Nile cruises especially in the autumn with passage observed from Suez.

**White winged Black Tern** *Chlidonias leucopterus* A common passage bird with small numbers wintering especially in Lake Qarun and the Nile Valley. Up to 100 were counted between Aswan and Luxor in 1990. Several flocks were seen on Lake Nasser in 1997 with one being as large as 300 birds feeding on a hatch of 'millions' of mainly dragonflies and damselflies. Immatures were seen on Lake Qarun in April 1997. Autumn counts at Lake Bardawil have had as many as 18,000 moving off the coast. Spring passage is observed from Suez.

**African Skimmer** *Rynchops flavirostris* Observed for a long time on Nile cruises with a maximum of 95 counted in 1986. It was in 1987 when birds were thought to be breeding but not until 1990, when it was proved, with 10-15 pairs found on a low sandy island north of Abu Simbel. There were 10 pairs in 1994. Khozam near Luxor is the farthest north the birds have been seen from the cruises, with up to 42 birds there. Two birds were seen at Aswan on 16th November 1996 and 3 at Kom Ombo on 19th February 1997 and 7 at Abu Simbel on 28th April 1997.

**Razorbill** *Alca torda* A very rare bird with no live birds being seen, only a washed-up corpse. No records exist for Israel.

**Lichtenstein's Sandgrouse** *Pterocles lichensteinii* A bird of rocky desert so limited to such areas as around Sinai and Gebel Elba. The best site to see them is at dusk when they come to drink at the Sharm el Sheikh sewage ponds where the highest count of 150 was on 13th October 1995. Up to 120 birds were found here in 1996 with a small count of 30 here on 17th May 1997. There is another site at Taba where 5 were seen coming to drink at their sewage ponds on 19th December 1995 and 6 were found the next day in Wadi Taba. Next door at Eilat there have been up to 140 seen in September 1993.

**Crowned Sandgrouse** *Pterocles coronatus* A common species of the Eastern Desert with a small number of records from west of the Nile. Water, again, is where to look for these species with Bir Beida near Quiser holding as many as 500 birds with 300 in 1995. They can be observed at several times of the day with small flocks coming into the wadi to feed as well as to drink. Dawn is the best time to observe this species at Sharm el Sheikh where up to 524 have been observed but more recently the figures have been more like 80 coming to drink at the sewage ponds on 7th and 9th May 1996. Nile cruises have seen several flocks coming down to drink on the Nile with the commonest site being near Qena where up to 20 birds were seen, and at Edfu. Six birds were found 100 kilometres from Aswan close to the Aswan-Abu Simbel road on 7th April 1997 and 32 between Farafra and Bahariyya on 7th February 1998.

**Spotted Sandgrouse** *Pterocles senegallus* This is the commonest breeding sandgrouse with records scattered over many regions other than mountainous areas. The largest population is found in North Sinai with several thousand regularly crossing the Israeli border to drink at the sewage at Nizzana Western Negev where up to 10,000 have been estimated at peak times. When this water dried up in 1985 the whole population moved over into Sinai to look for water. There was a mixed population of 5,000 here in 1995. A small number drink at Sharm el Sheikh where there were 7 birds on 7th May 1996. Nile cruises have seen several flocks of birds ranging from Meydum, Beni Hassan and Edfu with the best tourist site being at the camel station near Abu Simbel where up to 200 birds can be found feeding on hay seed brought to feed the camels. Small numbers have also been seen along the Cairo-Faiyum road. Up to 17 have been seen coming in to drink on Crocodile Island, Luxor.

**Chestnut bellied Sandgrouse** *Pterocles exustus* An extremely rare bird now with few recent records. It had occurred on Nile cruises with one record of a flock of 10 coming to drink on the Nile near Edfu in the 1970s.

**Black bellied Sandgrouse** *Pterocles orientalis* A bird of the North Sinai with up to 1,200 counted at Nizzana just over the border in Israel in December 1995 where a breeding population of 1,000 pairs was estimated in the 1980s. In recent years small numbers have been seen in Sinai with a record on 9th May 1997 of only 2 birds.

Rufous Bushchat

**Pin tailed Sandgrouse** *Pterocles alchata* Very few records exist for this nomadic species which in some years can number as many as 5,000+ in the Nizzana area of the Negev Desert in Israel. It is a species very much preferring rainfall areas in deserts where the spotted then moves out. There are records from Rafa and Wadi el Arish in northern Sinai.

**Rock Dove** *Columba livia* A species now likely to be overlooked by birders due to the domestic pigeon often breeding with it, especially around human settlements where domestic pigeons are used for meat and their droppings for fertiliser. The smaller darker race *schimperi* of this bird can be still found along the Nile and has been observed by cruises.

**Stock Dove** *Columba oenas* A winter visitor to mainly the northern Sinai with a single bird seen at Zaraniq on 26th December 1994. Flocks were seen flying along Wadi El Arish on 16th January 1995 which was the highest numbers counted in recent times in Egypt with over 100 birds present. A single bird was present at Sharm el Sheikh on 27th October 1995 with the farthest south record coming from Aswan Old Dam on 20th January 1990 when two birds were present. A single bird was at Gebel Asfar on 30th March 1995.

**Wood Pigeon** *Columba palumbus* A very rare bird in Egypt with the only records coming in the 1990s with a single bird shot 30 kilometres south of Alexandria and another single bird seen on 27th October 1995 at Sharm el Sheikh. The old records of birds from North Sinai were rejected even though it is a wintering bird of Israel where it can be found mainly in the north of the region but a small flock of 25 was found in the North West Negev on the Egyptian border as recently as 1977.

**African Collared/Pink headed Dove** *Streptopelia roseogrisea* A resident breeding bird of the Gebel Elba area with up to 9 birds found in 1997. The most accessible place to find this species is at Abu Simbel where as many as 10 birds can be found, as in January 1992, but smaller numbers are more likely. There were 7 on 21st November 1994 but only a single bird there in the spring of 1997. Other records come from Aswan where 2 possible birds were in the grounds of the Amun Sheraton on 18th January 1990 and 2 birds were present in the acacia at Shallal opposite Philae on 25th November 1994. Nile cruises have had birds as far north as Edfu while 3 were at Sharm el Sheikh on 17th April 1993.

**Collared Dove** *Streptopelia decaocto* An expanding species for the country with birds moving through Sinai from Israel. Records are spreading along the Nile and in the Delta region but surprisingly no birds were found in the markets in 1990 at Damietta and Port Said. This species was not commonly seen from Nile cruises with the first record only being in 1986 with 2 birds at Meydum and 2 were found at Kom Ombo in 1989 which was the farthest south record. Up to 45 birds were at Suez on 14th March 1998.

**Turtle Dove** *Streptopelia turtur* A very common breeding bird of agricultural areas and even moving into reed beds at Lake Manzala. Often found breeding alongside palm doves and easy to see at such sites as Crocodile Island and the Winter Palace Hotel gardens, Luxor. Many are eaten in Egypt and often found in markets up and down the country. One at Farafra by 7th February 1998.

**Rufous Turtle Dove** *Streptopelia orientalis* A rare bird of the area with the only known records coming from first of all Abu Simbel, where a single bird was found on 1st October 1990 with a second record from the Sinai Wildlife Centre at Coral Bay near Sharm el Sheikh, South Sinai on 18th October 1995. There are only two records for Israel both coming from Eilat with one on 4th September 1984 and another here on 27th April 1995.

**Palm Dove** *Streptopelia senegalensis* A very common species of both town and agricultural land with a sub species breeding at Gebel Elba. It is very commonly eaten in Egypt with up to 3,400 in the market at Damietta in the Delta in 1990.

**Namaqua Dove** *Oena capensis* A well sought-after bird with several records coming from both Abu Simbel and Sharm el Sheikh. The first proof of breeding came in 1997 with one pair and a second female present near Aswan on 30th April; Gebel Elba was always thought to be the first breeding area. Nile cruises have seen this bird as far north as Edfu with another record coming from Kom Ombo. A single bird has also been seen on Crocodile Island, Luxor. At Sharm el Sheikh. single birds were seen on 17th October 1997 and on 18th October 1997. A single bird was seen flying in off the sea at Zaraniq on 11th October 1994 and another single bird was at El Tor, South Sinai on 7th October 1996. A single was at Abu Simbel on 12th September 1996. These singles are dwarfed by the recent groups of birds found at Eilat in summer, with 47 being the highest.

**Ring necked Parakeet** *Psittacula krameri* A common breeding bird of several towns in Egypt with Cairo being the best for seeing them, especially around the sports fields on Gezira Island. Nile cruises have picked them up in Cairo with a maximum of 8 birds having been seen.

**Alexandrine Parakeet** *Psittacula eupatria* A pair of these birds were seen well in the gardens close to the Gezira sports fields on 10th and 11th April 1997. Like the ring necked parakeet, they may well form a wild population in Cairo, if not already doing so.

**Great spotted Cuckoo** *Clamator glandarius* A former common breeding bird of the Nile Valley but has declined rapidly maybe due to the past use of pesticides. Recent breeding records come from Aswan where at least two hooded crows were found with young cuckoos on 28th and 29th March 1996. Another area thought still to have breeding birds is El Arish. Passage is often found in spring with several records coming from Nile cruises with such sites as Ramesseum Rest House, Luxor, and near Qena. A single bird was at Wadi Kidd, Sinai on 20th February 1998. An estimated 500 pairs breed in Israel, using mainly hooded crow hosts but also brown necked raven and jay.

**Cuckoo** *Cuculus canorus* A scarce passage migrant with birds seen on Nile cruises as well as general records including a very late one from Luxor on 30th April 1991; one at Bir Beida on 18th April 1995 and another at Abu Simbel on 9th April 1997. It is a common breeding bird of northern Israel with several uncommon hosts including desert lark, rufous bushchat and Arabian babbler, with the commonest hosts being scrub warbler and long billed pipit.

**Senegal Coucal** *Centropus senegalensis* A breeding bird of the Nile Valley and Delta with many records coming from Nile cruises, especially from the Faiyum where it is also found. The main site for seeing this bird near Cairo is at Gebel Asfar where it is fairly common with few people missing it. The royal duck ponds of King Feluke below Dahshur is another site to see this bird. It also appears in the market at Damietta where up to 54 were found in 1990 as a table bird.

**Didric Cuckoo** *Chrysococcyx caprius* The first record for the Western Palearctic was from Cyprus in 1982 with the second and a more recent record of a long stayer at Eilat from 12th to 27th March 1994.

**Barn Owl** *Tyto alba* One of the best performances of the sound and light show at Karnak or Giza cannot be finished unless the sound of this bird is heard or an individual drifts by in the lights of the performance. It is regularly found in the temples and tombs of the Delta and Nile Valley with many cruises bumping into them as the sites are visited. Birds have been seen at Saqqara, Meydum, Tarnis and Abydos. Several new sightings came from the Delta in 1990 and a single bird was seen flying over hotel gardens at Naama Bay, Sharm el Sheikh on 1st February 1997. One bird was seen at Aswan on 25th March 1995.

**Striated Scops Owl** *Otus brucei* There are some very old records from Sinai for this species but, given the increasing number of records from Eilat new records for the country must be only around the corner. So a record from Wadi Taba of an over-wintering bird from November to February 1991/92 comes as no surprise. There were up to 6 birds ringed during January and March 1992 around Eilat.

**Scops Owl** *Otus scops* A fairly common migrant during the spring passage with fewer records in the autumn. Due to its nature of hiding away, many records must go amiss. A calling bird at Ain el Gedeirat, North Sinai may be the first proof of potential breeding in the country as it was found as late as 8th/9th May 1997. There is a breeding population in northern Israel. A single bird was found on 9th October 1994 at Quseima, North Sinai and one was at Marsa Matruh on 26th September 1997.

**Eagle Owl** *Bubo bubo* A widespread breeding bird often using the ancient temples and pyramids as roosting and breeding sites. The most famous pair is found at Saqqara where both adults and young have been seen regularly. A pair was seen at Giza Pyramids on 16th April 1995 and an adult and two chicks at the Step Pyramid at Saqqara on 17th April 1995 and one near Abu Simbel on 18th February 1997 where two pairs were found on 10th May 1997. A single bird was photographed in Wadi Allaqui in May 1997. Nile cruises have seen them at several sites, with the temple at Edfu being a new site for them.

**Little Owl** *Athene noctua* A fairly common breeding bird of the ancient monuments and cultivated land. Often seen at such sites as Giza, Hathor, Abydos, Tarnis, Luxor and Edfu. It is not so common in Sinai with records from Lake Bardawil and Rafa. A single bird was at Marsa Sagfre, Red Sea on 6th December 1997.

**Hume's Tawny Owl** *Strix butleri* A scarce breeding bird of the southern Sinai mountains and wadis with calling birds heard at St. Catherine's with up to 2 birds found here in March 1997 and others commonly located at Wadi Feiran. There has been a pair heard calling at Wadi Mukattab and others at Gebel Serbal, Wadi Watir, west of Nuweiba, and Wadi Nasb. There is an estimated 200 pairs in Israel. A new confirmed site is around Gebel Elba where up to 4 birds were found calling in 1997. This is the farthest south this species has been found in Africa. A bird was seen north of Aswan on 23rd December 1993 and again on 2nd January 1994 with a tawny owl type thought to be this species seen at Abu Sir on 31st March 1995. There are no other records for this species in this part of the Eastern Desert.

**Long eared Owl** *Asio otus* An uncommon winter visitor with most records coming from the Delta with others spread as far south as Abu Simbel. Single records came from Wadi el Arish, Suez and Shakshuk, Lake Qarun in 1990. Birds have also been found wintering in Cairo from 28th February to 5th March on the sports fields, and possible breeding was suspected here. A recently-killed bird was seen in Alexandria on 18th December 1992. There was a single spring record of a bird roosting in acacia at Shallal near Aswan on 6th April 1997.

**Short eared Owl** *Asio flammeus* An uncommon winter visitor with records mainly along the northern coast but inland records from Lake Qarun and Wadi el Natrun. Nile cruises have caught up with this bird at Lake Qarun and near Beni Suef and there were two late spring records from Lake Manzala on 19th and 21st May 1990.

**Nubian Nightjar** *Caprimulgus nubicus* This is thought to be a breeding bird of the Gebel Elba region but unfortunately there are no recent records. This species has declined rapidly in southern Israel due to the bird's lack of respect for motor vehicles. At night it sits on the road and waits for the vehicle to pass! Small numbers were observed at Sharm el Sheikh when this population was strong. No recent records exist.

**Nightjar** *Caprimulgus europaeus* An uncommon migrant with most of the records from the north coast of Sinai especially at Lake Bardawil where up to 472 have been counted in the autumn. A single bird was found at El Til on 16th March 1986, being the only record of this species during Nile cruises

**Egyptian Nightjar** *Caprimulgus aegyptius* A bird well-looked for by visiting birders but many go home without it on their lists due to the bird blending in so well with the background. Nile cruises have been fortunate to have a site to visit at El Amarna where up to 15 have been counted in recent years. Other sites include Kom Ombo and Abu Simbel. Close to Cairo the Abu Sir pyramids can produce up to 10 birds in the north west corner of the escarpment. Even hiring horses has been suggested as the best way to see them as the birds can be flushed by them. Birds have also been seen around the temples at Abu Simbel with up to 10 birds counted here on 11th September 1996 and 5 on 16th February 1997. There were 2 birds at Sharm el Sheikh on 27th/28th March 1996 and 2 south of Marsa Alam on 9th December 1997.

**Swift** *Apus apus* A common passage bird both in the autumn and the spring with records coming from around the country. This species was classed as uncommon on Nile cruises with the bulk of the records referring to pallid swift. First spring records were recorded at Port Said on 8th March 1990 with a late record on 8th June at Saqqara and Gebel Asfar. General passage has been watched at Suez where the peak of movement was on 6th May with 120 birds. A large population is found breeding in northern Israel with 10,000 + pairs with at least 4,000 pairs in Jerusalem alone. The nearest breeding site to Egypt will be in the Negev Desert.

**Pallid Swift** *Apus pallidus* This is a common breeding bird of the towns north of Luxor with smaller numbers wintering with birds found along the Nile and even in the Delta in 1990 and resident birds found at some of the southern oases like Dakhla. Large flocks were observed in Cairo from 8th February 1997 up to 14th December 1997. This bird was commonly seen on Nile cruises with migrants seen south of Luxor. Breeding birds of Israel are mainly found on cliffs and uninhabited areas, not in towns. A small population is thought to breed like this high in the mountains around St. Catherine's Monastery in Sinai.

**Alpine Swift** *Apus melba* This is an uncommon migrant with a few records each year with the bulk of them coming from Suez. This bird was thought to breed on Sinai but that was only proven in the 1990s with a colony found at Sad al Rawafa in the north. Several thousand breed in the mountains of Israel where they arrive as early as January which may account for some of the winter records in Egypt. One or two birds have been seen on Nile cruises but the only recent record for the Red Sea comes from Ain Sukhna on 25th March 1994. A single bird was seen at Abu Simbel on 26th March 1995.

**Little Swift** *Apus affinus* This is a very rare passage migrant with some records in the winter in Sinai. There was a single bird at Sharm el Sheikh on 21st and 22nd November 1991. A breeding population is found in north east Israel of several hundred pairs. Records from Eilat show it was a spring migrant rather than an autumn one.

**Palm Swift** *Cypsiurus Parvus* This species has been added to the list of birds losing their habitat in which they were found before the Aswan Dam. The records suggest a decline well before the dam was built but once the habitat has gone there is little chance of a return. They are still found in Sudan.

**White breasted Kingfisher** *Halcyon smyrnensis* This is one of the birds most people want to see on a trip to the country. A small population now breeds in the southern part of the Delta with the best site to see them at Gebel Asfa where up to 3 pairs have been recorded since 1985, with 12 pairs found in the area in 1990. The other known breeding site is at Rafa, North Sinai where two pairs were found on 26th April 1992. In recent years birds have also been turning up at Sharm el Sheikh where 2 were found on 13th October staying until at least 27th October 1995 with one wintering here in 1994/95 and another bird here from 23rd September to 6th October 1996 and February 1998.

**Common Kingfisher** *Alcedo Atthis* This bird is thought to be a rare breeding bird but a common migrant with an amazing count of 1,195 passing Zaraniq, travelling along the Mediterranean coast in the autumn. Ringing returns have recovered birds from both Bulgaria and the Rjazan region of Russia. It is a common winter bird of the Delta and the Nile Valley with many records coming from Nile cruises usually of that bright blue flash as it moves across the water. Records come from as far as Aswan but a single bird was seen at Abu Simbel on 9th April 1997. It travels all down the Red Sea coast using, especially, the mangrove areas, and is also seen in South Sinai. A single bird was found along the Corniche at Marsa Matruh on 26th September 1997.

**Pied Kingfisher** *Ceryle rudis* One of the favourite birds of a Nile cruise with up to 630 birds counted in a short stretch of river. The birds have moved farther north along the river since the mid-1980s when a cruise would not see the birds until Sohag, with birds being regularly found now in Cairo and on Lake Qarun. There also seems to be an expansion in the Delta Lakes with records well inland from their normal coastal area. Expansion of cultivated land, using canals, will also push the birds into the Western Desert (11 at Dakhla Oasis in February 1998) and Sinai. A rare record comes from South Sinai on 4th January 1995 with a single bird at Shora el Mantaqa and at Abu Simbel in February 1998.

**Little green Bee-eater** *Merops orientalis* This is a common breeding bird of the Nile Valley and Delta area with also a population around the Faiyum and Wadi el Natrun. The birds can easily be seen in the centre of Cairo with the Giza Zoo and other parks offering good views. Nile cruises have often had the birds using the boat as a perch from which to fly out and catch insects, and a total of 296 have been counted along the river and at monuments. Small numbers are seen along the Suez and a new population, of the Levantine race found in Israel, has moved over the border and is now breeding in South Sinai at sites such as Taba and along the Taba-Nuqb road.

**Blue cheeked Bee-eater** *Merops superciliosus* This is a breeding bird of the Nile Delta and Upper Nile with small numbers nesting elsewhere. It is common on passage with large numbers recorded, especially along the Nile where it was commonly seen by cruise members at such sites as Karnak, Aswan, Philae and Balliana where they were seen with the two other species of bee-eater in the same area. Small numbers now breed in North Sinai and a small population is found next-door in Israel, breeding mainly to the east of the country.

**Bee-eater** *Merops apiaster* A very common passage bird with recent records of breeding along the Sinai coast with birds often mixing with blue cheeked Bee-eaters to nest. The birds do nest just over the border in Israel especially along the coast down to the northern Negev. The total population for Israel here is estimated at up to 5,000 pairs. Nile cruises have commonly seen this species, with counts as high as 745 birds for a trip with sightings often around temples and the river bank. Passage has been observed from Suez where an old record of 5,500 birds passed in 3 hours but a count of 928 birds inland at Faiyum in 1990 was also exceptional.

**Roller** *Coracias garrulus* Small numbers are found on migration both in the spring and in the autumn. Nile cruises have seen several birds over the years with more records from Aswan than anywhere else with 4 there on 16th September 1993. Other records have come from the Delta areas and the North Sinai coast with Red Sea records from even Gifton Island, off Hurghada. Up to a thousand pairs are found breeding in mainly northern Israel but with some records from the northern Negev.

**Abyssinian Roller** *Coracias abyssinicus* With an increasing number of birders making the pilgrimage to Abu Simbel to find African species in the Western Palearctic, new records are bound to arise and 2 recent records on 29th September 1995 and 7th May 1997 make this the site for finding this species. This brings the total to 5 records with 3 of them from here. There are no records for Israel of this species.

**Hoopoe** *Upupa epops* This is a very common breeding bird of the Nile Valley and Delta with new sites coming from the grounds of tourist development around the country due to the greening of areas. It is also a common passage migrant with many records from Nile cruises including a count of 266 seen along the cruise. Passage has been watched in the autumn from Zaraniq in North Sinai with up to 438 coming in off the sea over several days.

**Wryneck** *Jynx torquilla* This is a common migrant with an increase of records from, especially, along the Red Sea where the new hotel complexes are used as stop-over sites with birds recorded, especially in the spring with records from March until into May. Nile cruises have added to this list with birds seen from the Valley of the Kings, Crocodile Island and Aswan. At least 50 birds were recorded from Wadi el Natron between 13th and 15th April 1990 and other birds were found at Lake Qarun, Gifton Island, Hurghada and Suez.

**Syrian Woodpecker** *Dendrocopus syriacus* This is one of the last species thought to be in Egypt due to the lack of native trees but this species has found the local palms, and the population has come over the border into North Sinai where birds have bred since the early 1970s. A new nest-hole was found at El Arish in 1996 and birds are regularly seen at Rafa.

**Black crowned Finch Lark** *Eremopterix nigriceps* This species is mainly found around the coast of the south east corner of the Red Sea with 6 singing males found at Abu Ramad on 1st April 1997. There was a nomadic movement of birds into Sinai during the 1970s, with several records coming from near Sharm el Sheikh and again near Eilat in the 1980s.

**Chestnut headed Sparrow Lark** *Eremopterix signata* A single record of this African species at Eilat on 1st May 1983.

**Dunn's Lark** *Eremalauda dunni* This nomadic species has been seen in recent years in North Sinai where 7 birds were found 45 kilometres south of El Arish on 29th August 1991 but none were seen in Israel that year after a large influx during 1989. Up to 6 males were displaying on 27th April 1992 and 3 found by Wadi Girafi on 21st December 1994.

**Bar tailed Desert Lark** *Ammomanes cincturus* This is a locally common breeding bird of the sandy deserts especially away from vegetation. The majority of records come from the Eastern Desert and Sinai, moving west along the Mediterranean coast to the Libyan border. Recent records have come from the Red Sea coast at such sites as Ain Suchna, on the road to St. Paul's Monastery, Bir Beida, Marsa Alam to Edfu road and Wadi Gemal. West of Suez, birds have been seen in Wadi Hagul and west of Cairo on the Faiyum road. It was a bird normally missed by Nile cruises with the only record coming from Beni Hassan. The highest count for this species comes from south of Sharm el Sheikh Airport with 108 birds in 1981.

**Desert Lark** *Ammomanes deserti* This bird is found in the more mountainous areas right down to the edge of cultivation along the Nile where cruises have generally found the species quite easily, especially from the monuments. One bird in April 1997 was acting like a sparrow flying down within 2 feet of people and taking bread back to its nest site near the harbour going out to Philae. It is commonly seen from the Cairo-Faiyum road and at the pyramids of Giza, Saqqara and Dahshur as well as sites such as Hathor, Beni Hassan, El Amarna and the Valley of the Kings.

**Hoopoe Lark** *Alaemon alaudipes* A local breeding bird of the low sandy areas especially away from mountains. Many records have come from Nile cruises with Saqqara one of the easiest sites to find this bird where up to 3 have been found. Also recorded from cruises at Beni Hassan, El Amarna, the Valley of the Kings and along the Cairo-Faiyum road. Singing birds were also seen at the newly-opened pyramid of Dahshur in 1997. Other records come from Dakhla near Mut in the Western Desert, the Cairo-Ismailiya road and Wadi Hagul.

**Dupont's Lark** *Chersophilus duponti* Recent observations have found a small population breeding along the Marsah Matruh-Sallum road with up to 8 birds seen during March 1994. Small numbers were also found 52 kilometres south of Marsah Matruh on the Siwa road on 28$^{th}$ September and 2$^{nd}$ October 1995. One was heard 40 kilometres east of Marsah Matruh and several found in June 1995 west and south of Matruh with 4 birds present in May 1997, also south of here.

**Thick billed Lark** *Rhamphocoris clotbey* A flock of 15 were found 52 kilometres south of Marsah Matruh on 28$^{th}$ September 1994. These represented the first records since 1969 with two on 12$^{th}$ June and three on 14$^{th}$ June 1995. Young were finally seen on the wing in the same month, making this the first confirmed breeding record for the country as a female with a brood patch had been found before, as long ago as 1922. Singles have been found in Sinai in 1968 and 1969 with a higher count of 15 in Moon Valley on the border near Eilat in 1978 and two flocks of 9 and 5 on the beach of Eilat in 1979 and 4 near here in 1989.

**Calandra Lark** *Melanocorypha calandra* This is both a passage migrant as well as a winter visitor, mainly to the western Mediterranean coast with small numbers feeding in the cultivated fields here. A flock of 8 was found on Ras el Hekma on 1$^{st}$ March 1997 feeding with skylarks and crested larks. There were 3 birds near Ismailiya on 15$^{th}$ March 1997. It is a rare bird inland with old records from Cairo and Giza but a single bird was found on 30$^{th}$ March 1997 close to Wadi el Rayan feeding in the Western Desert close to 3 small bushes. There are a few records from along the Red Sea and also from Sinai where it has occasionally bred east of Rafa, which must be birds coming over from Israel where up to 6,000 pairs breed.

**Bimaculated Lark** *Melanocorypha bimaculata* An uncommon migrant with most records coming from along the Red Sea where hundreds of birds may pass in late March. Possible winter records come from St. Catherine's where a single bird was in November 1984; 5 were found west of Ismailiya on 3$^{rd}$ January 1990 and from Sharm el Sheikh on 23$^{rd}$ November 1991 and from Safaga on 3$^{rd}$ October 1997. Feeding flocks of up to 1,000 birds can be found near Eilat in the spring.

**Short toed Lark** *Calandrella cinerea* A very common passage migrant with records from many parts of the country especially in the spring. Large flocks are often seen crossing the Cairo-Faiyum road and also moving up the Red Sea coast where several thousand have been counted. In the autumn 5,623 have been counted in September moving south over Zaraniq. The highest feeding flocks have been counted just over the border at Nizzana where up to 10,000 birds were counted in March 1996. There is also a small breeding population here but the only recent possible breeding birds in Egypt have occurred at Wadi el Natrun where a single bird was singing to its mate on 17th May 1990. This bird has been seen many times on Nile cruises, especially in the spring.

**Hume's Short toed Lark** *Calandrella acutirostris* A single bird was found at Eilat on 4th February 1986.

**Lesser Short toed Lark** *Calandrella rufescens* A local breeding bird with two distinct forms with one breeding in the Western Desert and the other found mainly along the Delta coast where it is a common wintering bird with a flock up to 150 birds found at Lake Manzala in January 1990 but only 5 singing birds found here in the spring. The Sinai holds another small breeding population with birds drifting over from the Israeli side where up to 1,000 pairs breed, often alongside short toed lark.

**Crested Lark** *Galerida cristata* This is a great favourite on Nile cruises as it often sits around the shore for everyone to view when the boat is moored up. It is a common breeding bird of the Nile Valley and Delta with birds spreading into new cultivation schemes, both in the Western Desert and Sinai. It is very much a coastal bird along the Mediterranean and Sinai coasts with only a handful of records inland away from habitation.

**Thekla Lark** *Galerida theklae* This is a rare breeding bird of the Western Mediterranean coast from Salum to the Libyan border. A pair found 10 kilometres east of Salum on 11th September 1993 was the first confirmed record in 60 years. Three males were displaying on 15th March 1994 and 2 birds were found here on 29th September 1994. A visit to the area found several singing males in June 1995 and 1997.

**Wood Lark** *Lullula aborea* This is an uncommon migrant with records coming from the Mediterranean coast and along the Red Sea with a few inland records from Sinai and a recent one from Wadi el Natrun on 15th March 1990. Montaza in Alexandria held 5 birds on 14th November 1992. Spring passage has also been recorded at Suez with a single bird moving north here on 12th March 1990. This species is more a northern visitor in Israel, with few records at Eilat.

**Skylark** *Alauda arvensis* A common winter visitor to the northern half of Egypt with many found along the coastal roads west of Alexandria. Recent counts have found up to 10,000 just over the border at Nizzana in the spring. Two birds were found at Garf Hessein, Lake Nasser on 29th January 1990 and inland movement was seen at Wadi el Natrun on 15th March 1990 and also at Dahshur on 29th March 1997. Birds were heard calling and singing in the El Alamain cemetery on 1st March 1997. This bird is rare at Sharm el Sheikh with one wintering in 1994/95.

**Oriental Skylark** *Alauda gulgula* With the increasing number of records at Eilat where birds have been observed from the early 1980s, it was only time before the first record for Egypt which came with a single bird on 14th October 1991 at Sharm el Sheikh. This was quickly followed by other records all from the same location with another single on 13th/14th October and 2 present on 20th October 1995. Another likely site for this bird is over the border from Nizzana where a single was found in 1995 and 2 birds wintered in 1996/97.

**Temminck's horned Lark** *Eremophila bilopa* This is a breeding bird of the northern part of Egypt with small numbers seen at several locations including 12 near Bir el Nuss at Matruh on 30th December 1989, 7 at Wadi el Natrun on 15th March 1990 and a single along the Cairo-Faiyum road. There were two birds at the 52 kilometre mark south of Matruh on 27th September 1997. On the Sinai it is again found in the north with 5 seen at Wadi el Arish on 13th January 1990 and an old record of a single bird at Ras Mohammed. It seems to be a bird that many birders miss in Egypt.

Great spotted Cuckoo

Masked Shrike

Nile Valley Sunbird

White winged Black Tern

Waiting for the plane, Abu Simbel-Masked Shrike, Hoopoe, Montagu's Harrier.

Spur winged Plover

African pied Wagtail

Kittlitz's Plover

Crowned Sandgrouse

Lake Nasser

Abu Simbel Temple

Pelicans on the Lake

Scorpion

Solid rock of the Desert

The lake's many outcrops

Lake Nasser

Sacred Ibis

White Pelican

Yellow billed Storks/Pink backed Pelicans

Long eared Owl

Cultivation by the Lake

Black headed Heron

**Brown throated Sand Martin** *Riparia paludicola* This African species is found south of the Egyptian border in Sudan but surprisingly only 2 records are found to date, both well away from this border. A single bird was seen at St. Catherine's in Sinai in May 1985 and 2 were seen travelling with a large movement of swallows and martins at Suez on 20th April 1990. The Sinai record is interesting as in the same month a single bird was caught and ringed at Eilat on 6th May which was the first record for Israel. The latest record from Eilat came on 10th and 11th November 1996.

**Sand Martin** *Riparia riparia* A very common breeding bird of the Nile with as many as several hundred thousand pairs with up to 157,759 birds counted on a small section of a Nile cruise in 1985, with an estimated million pairs present. Wintering and early returning birds are seen in January with several hundred present at Crocodile Island on 26th January 1997 with birds also singing. Large numbers have also been seen over the Delta Lakes with 10,000 over Lake Manzala in May 1990.

**Banded Martin** *Riparia cincta* A record of this African species was from Elephantine Island, Aswan on 15th November 1988. There are no records from Israel or the rest of the Western Palearctic for this species but there was one in North Yemen in March 1982.

**Pale Crag Martin or African Rock Martin** *Hirundo obsoleta* This is a common breeding bird of the mountain and rocky areas of the Nile Valley, Eastern Desert and Sinai with the first birds seen on a Nile cruise usually from Beni Hassan and from there up to Aswan and again at Abu Simbel. The koppie rocks around Lake Nasser hold several pairs with nests inside holes as well as being glued to the underside of overhangs. Eggs were found here on 24th January 1997. One amazing nest was found underground with the bird flying out of an open drain hole cover. Peak movement for this bird at Eilat is in February and November.

**Crag Martin** *Hirundo rupestris* With strong light this species can be confused with the pale crag martin and several records have not been confirmed. There is a small wintering population around the Upper Nile with Aswan being the best place to see them. A small number of records come from Nile cruises with other records from Luxor and migration from Suez up until the end of April. The latest spring record came from Aswan on 1st May 1991.

**Swallow** *Hirando rustica* The distinctive Egyptian form with its red breast is another favourite on Nile cruises as so many of the guests have seen the light bellied form back in their native country. This breeding population is found both in the Delta and along the Nile with a small population found down the Suez Canal. The pale form has bred occasionally from Port Said back along the coast to the Israeli border where it is a common breeding bird, with around 10,000 breeding pairs.

**Ethiopian Swallow** *Hirundo aethiopica* This African species was first seen in Israel on 22nd and 23rd March 1991.

**Wire tailed Swallow** *Hirundo smithii* This African species was first seen on 30th March 1995 north of Wadi Hagul off the Suez-Cairo road. This is the first known record for the Western Palearctic.

**Red rumped Swallow** *Hirundo daurica* This is a common migrant with smaller numbers than swallows seen together. Several records have come from Nile cruises and birds have been seen in the spring from the beginning of March with autumn records until November. One summer record was of 2 birds at St. Catherine's, Sinai on 12th June 1990. An estimated 15,000 pairs breed mainly in northern Israel but some are found in the North West Negev.

**House Martin** *Delichon urbica* This is a common passage bird, especially in the spring with migration lasting from March to early June. Passage has been observed at Suez with up to 380 birds moving through in April. A small number of birds have wintered, with records from Sharm el Sheikh and Edfu. A population of up to 2,000 pairs breed in northern Israel.

**Richard's Pipit** *Anthus richardi* This is one of many species which are gaining records due to increased observers on the ground, with Sharm el Sheikh the site where the bird can more easily be found on migration and when wintering. The most found so far has been 11 birds here from 13th-20th October 1995. There are both spring and autumn records. Further spring records include one on 17th and again on 27/28th March 1996 with an autumn peak of 6 birds in October 1996. Away from here there was a single bird at Taba on 4th October 1996 and inland, a single bird by the Dahshur pyramids on 29th March 1997. Wintering birds in Israel come right up to the Egyptian border at Nizzana and Eilat, so try the Taba hotels!

**Blyth's Pipit** *Anthus godlewskii* The only record comes from Eilat with a single bird in fields on 7th November and again on 16th November 1987.

**Tawny Pipit** *Anthus campestris* A common migrant in smaller numbers than other pipits, with records from around the country. Suez has recorded up to 124 birds with most records of 1-5 birds. This was the case in 1997 with birds seen in the spring from Lake Qarun, Dahshur, Ras Zafarana, Hurghada, Crocodile Island, Aswan and Abu Simbel, with the highest count being 3 at most locations. It has been seen by several Nile cruises in many locations and there are a number of wintering records especially from the Delta Lakes, with 6 at Lake Maryut and 5 seen at Wadi el Natrun in 1990. There is a wintering population on the border at Nizzana where the birds are often seen feeding with Richard's pipits.

**Long billed Pipit** *Anthus similis* The first records of this species came when 2 were found on Gifton Island off Hurghada on 26th September 1996. Then with a visit to Gebel Elba up to 12 birds were found 1st April 1997 which may be the first indication that this species is breeding in the country. The finders of these birds suspect that the species has moved north from southern Sudan not south from Israel where it also breeds.

**Olive backed Pipit** *Anthus hodgsoni* Another species so far only found from Eilat where up to 15 can be found in the autumn, with birds even wintering farther north. It is often found with tree pipits which are a common migrant in Sinai and along the Red Sea coast so it may soon be found here.

**Tree Pipit** *Anthus trivialis* A common passage bird in particular in the spring with large numbers noted especially along the Red Sea where passage has been watched from Suez. The bird is often found with red throated pipit in this area. Up to 40 birds were seen in the Faiyum on 15th April 1990 with 26 caught and ringed at Shakshuk, Lake Qarun on 26th April 1990.

**Meadow Pipit** *Anthus pratensis* A common migrant with a winter population especially along the western Mediterranean coast where there is noted a lack of red throated pipits. It was commonly found on Nile cruises especially as the groups walked from the boat, through the agricultural land to the ancient sites.

**Red throated Pipit** *Anthus cervinus* A very common migrant and winter visitor with more birds found to the east of the country than to the west. It is now a common species to see around hotels such as the Hurghada complex, Crocodile Island, Luxor and Abu Simbel with records from Cairo coming from the Gezira sports centre. Nile cruises often bumped into this bird, with Kom Ombo always a special place to see this species.

**Water Pipit** *Anthus spinoletta* A wintering bird of the Delta Lakes with 100s seen in 1990 with perhaps 1,000s present, especially around eastern Lake Manzala. It is also a passage bird with large numbers seen passing Suez where a roost of 950 birds was found in the autumn of 1981. Other records come from Wadi el Natrun where 46 were counted in 1990; also Lake Qarun, Lake Bardawil, Aswan and Abu Simbel as well as Western Desert Oases.

**Buff bellied Pipit** *Anthus rubescens japonicus* This bird was discovered wintering at Eilat only in 1981/82 when at first it was thought to be a 'short winged water pipit'. New work indicated the species and since then from 4 to 60 birds have wintered. The first in Egypt were found at Lake Qarun in March 1993 when 2 birds were present and a single bird was found at Sharm el Sheikh on 24th December 1995 with 2 more at Zaraniq on 21st March 1997. Who has checked all the water pipits on Lake Manzala for this species?

**Yellow Wagtail** *Motacilla flava* Egypt has its own breeding population of this species which has so many forms and plumages the visitor has to look closely to see what is in front of them. The local breeding population is found along the Nile and in the Delta area including Wadi el Natrun and the Faiyum. Winter concentrations include flocks of 400 around Lake Manzala in January and family flocks of 250 at Ismailiya by the beginning of May. General migration of the other forms have included 1,490 counted on one day in September at Zaraniq and Suez, passage being observed from September into November. The other forms include records for the following:- Flava, Feldegg, Thunbergi, Beema and Cinereocapilla. Some hybridised forms can also occur. A ringed bird from near Cairo was found in Finland 4 years later. The late W.B. Alexander, the first Director of the Edward Grey Institute of Field Ornithology, is said to have stepped off a cruise ship at Suez and seen 13 of the 12 races of yellow wagtail!

**Citrine Wagtail** *Motacilla citreola* This is another bird which would benefit from an increase in birdwatchers as records from Eilat can average 100 birds in the spring and 50 in the autumn. So far, few records exist here with a single bird at Lake Bardawil in 1985 followed by a male at Lake Manzala on 10th April 1990. There was one at Hurghada on 18th March 1992, one on 17th/18th October 1992 at Sharm el Sheikh, a single bird at Sheikh Zweid, North Sinai on 30th March 1995 and another at Nuweiba on 30th March 1996 with another at Sharm el Sheikh on 6th May 1996. Abu Simbel had a single bird on 9th September 1996 and again on 9th April 1997.

**Grey Wagtail** *Motacilla cinerea* A winter visitor in small numbers especially along the Nile where several cruises have picked up this bird. Records have come from many sites including the Egyptian Museum, Cairo where birds have been attracted down to the fountain in front of the museum. They have also been seen at Faiyum, Suez, Hurghada and down to Aswan on the Nile. Passage has been seen from Suez in both spring and autumn with early birds seen by 3rd October.

**White Wagtail** *Motacilla alba* A very common passage and winter visitor to all areas of cultivation but even to Bedouin camps, well out into the desert. The birds appear to come from Scandinavia with most ringing returns coming from here but a longer ringing programme is needed to confirm this. Up to 100,000 have been seen coming to roosts, even in a Cairo suburb, and an old record of migration in late March estimated a million birds flying out over the Mediterranean, with thousands dead next day on the ship which witnessed this movement.

**African Pied Wagtail** *Motacilla aguimp* This is a former bird of the Nile Valley around Aswan which disappeared when the big dam was built. In the early 1980s records started to come back, with birds especially around Abu Simbel with 5 pairs found on 2nd May 1994 which was the first confirmed breeding for this species in the Western Palearctic. Nile cruises have occasionally bumped into this species if they took the Abu Simbel excursion but many more birders can miss it if they do not take the boat trip out to their breeding islands. There were 8 birds present on 11th September 1996, and a pair on 8th April 1997 which seemed to be going to nest in an old oil drum on the side of the lake. This may have been the first mainland nest site recorded.

**Common Bulbul** *Pycnonotus barbatus* This is a resident species of the Nile Valley and Delta with breeding birds also found at Faiyum, Wadi el Natrun and the Suez Canal down to Ein Sukhna. It is extremely rare in Sinai and with large amounts of development of new towns and agricultural land taking place it will be interesting to see if the bird moves east towards the yellow vented form. It is very common in towns like Cairo where it can be seen even in street trees but especially if you can find gardens. A separate colonised area is south, in Gebel Elba where over 500 kilometres of desert means the species is undisturbed.

**Yellow vented Bulbul** *Pycnonotus xanthopygos* This is a widespread species in towns and wadis in Sinai with birds found along the Mediterranean as far as Romani. Newly-developed tourist centres may well spread the species along the coastal areas and may be up to the border of the common bulbul's range. There are very few records of hybridisation between the two species.

**White cheeked Bulbul** *Pycnonotus leucogenys* This species was new to Eilat in 1988 and may be spreading its range with several records coming from here in recent years. These also could be records of escaped caged birds! The last records were of two birds here from 14th March into April 1994.

**Grey Hypocolius** *Hypocolius ampelinus* A male was first seen at Abu Simbel on 16th November and a female joined it on 8th December 1995. The only other record for this species is way back in 1938. Both of these birds seem to be wintering and it is yet another exciting find from this site. Only 4 were seen at Eilat between 1987 and 1989 with the last record being a pair which stayed from 21st November to 4th February 1990. Another bird was found in 1997 at Eilat which was easily photographed eating water melon!

**Wren** *Troglodytes troglodytes* This is a bird not normally found in Egypt but on 1st February 1997 a single bird was observed moving along vegetation at the hotel at Al Alamein on the Mediterranean coast but was not watched closely, mainly because the observer thought it was just a 'wren'. There is a small breeding population in the north of Israel with two dates from Eilat in 1982.

**Dunnock** *Prunella modularis* A rare winter visitor to the Mediterranean coast with 2 at Port Said in 1990, one at Montazah, Alexandria on 25th December 1992 and one on 22nd January 1993, west of Alexandria. There was a single bird at Port Said on 28th January 1995. There are a small number of records from elsewhere with a single bird at Suez in March probably on migration. A single record comes from St. Catherine's in Sinai and one from Giza, Cairo. Birds have been observed at both Eilat and Nizzana but not annually, with over a 1,000 birds wintering in the north of the country.

**Rufous Bushchat** *Cercotrichas galactotes* A breeding bird of the Nile Valley and Delta area with other records coming from Wadi el Natrun where 8 singing birds were heard on 15th May 1990; S Saqqara held 2 singing birds in June 1990; and also Faiyum, Edfu and Kom Ombo. The guaranteed site for this species is in the grounds of the Movenpick Hotel on Crocodile Island, Luxor where the birds are so used to people they walk around in front of you. On Sinai they are mainly found along the Mediterranean coast but with so much development of hotels, they may well expand with this.

**Black Bushchat** *Cercotrichas podobe* A newly-found species visiting Eilat with the first record in 1981 with an average of 5 annually since and a possible breeding bird of the Arava Valley. The bird was thought to be a resident in Gebel

Elba due to a population on the Sudanese Red Sea coast but recent visits have not found the bird. There are two recent records from Sinai with one at Nuweiba in 1993 and one at Naama Bay on 31st January 1997.

**Robin** *Erithacus rubecula* A winter visitor especially in the north of the country along the coast with few records inland. A small number of records come from the Suez area with very few south of Cairo. There was a single bird at the Agricultural Museum on 14th February and another outside the National Conference Centre on 16th February 1997. Several birds were seen along the west coast on 1st March 1997 as far as Marsa Matruh with a reuturning bird here on 26th September 1997. Records have also come from Port Said, St. Catherine's, Ain Sukhna and Faiyum.

**Thrush Nightingale** *Luscinia luscinia* A fairly common migrant in the autumn but with fewer records in the spring. The hotel complexes of Hurghada are turning up several records especially from the Sheraton Hotel where birds have been seen up to 8th May. Sinai records are chiefly from the north of the peninsula around Lake Bardawil but also from the high mountains. Eilat has higher numbers in the spring and with new hotels opening up in the Sinai more records could come from these in the future. Nile cruises have turned up this bird at El Amarna and Philae. Ringing has shown that many of these birds pass through the country from Scandinavia.

**Nightingale** *Luscinia megarhynchos* A common migrant mainly in the spring with records from around the country especially from Cairo at the Gezira Sports Centre, Faiyum and Suez. Several records come from the hotel complexes from Sharm el Sheikh to Hurghada. The Nile cruises have found this species at sites such as Aswan, Kom Ombo and Luxor, all in the spring.

**Bluethroat** *Luscinia svecica* This is a very common winter visitor to areas of cultivation and reed beds. It can be found in the middle of Cairo at the Gezira Sports Centre, Giza Zoo or behind the Agricultural Museum. It was found by many of the Nile cruises from Lake Qarun right down to Abu Simbel where it feeds along the trees as you approach the temples. Coastal records again come from the hotel complexes at both Hurghada and Sharm el Sheikh.

**Siberian Rubythroat** *Luscinia calliope* A first winter female was found at Hurghada from 9th to 12th November 1993. This could be the first record for Egypt and again shows the benefit of these hotels for offering migrants a green stop-over site. There are no accepted records in Israel yet.

**White throated Robin** *Irania gutturalis* A rare migrant with two records from Sinai with a single bird near Abu Zenima, South Sinai on 30th May 1984 and another at El Arish, North Sinai on 3rd May 1990. There are several records from Eilat with a maximum of 39 birds in 1984 but not seen every year and up to 10 pairs breed in the far north of the country.

**Black Redstart** *Phoenicurus ochruros* A fairly common winter resident in the north of the country with most birds departing by late March with 8th April the latest record in the spring. Nile cruises have seen this bird at such sites as Giza, Saqqara and Meydum pyramids and Aswan which is the farthest south for the species along the Nile but birds have also been seen at Gebel Elba in the far south east corner of the country. A general movement of birds was noticed on 1st March 1997 on the western coast with birds located at Marsa Matruh, Al Alamein, El Alamein and the outskirts of Alexandria.

**Redstart** *Phoenicurus phoenicurus* This is a fairly common migrant with occasional birds wintering even in Gezira Sports Centre. It is a bird found on Nile cruises from several locations especially the Winter Garden Palace at Luxor and the Cataract Hotel, Aswan. Spring migration can start as early as February with records from the hotels at Hurghada and Sharm el Sheikh.

**Blackstart** *Cercomela melanura* This is a breeding bird of mainly the south and east Sinai with more found in the mountains with another breeding population found at Gebel Elba. Away from these sites birds have been found by Nile cruises at Saqqara, Beni Hassan, Philae and Abu Simbel. A recent record came from here on 6th May 1997 and it is thought that a new breeding population may exist around the shores of Lake Nasser.

**Whinchat** *Saxicola rubetra* This is a fairly common migrant moving through the country both in spring and autumn. Few records have been found on Nile cruises but in April 1997 this bird was seen in the Western Desert between Aswan and Abu Simbel through to the centre of Cairo. It has been found in the grounds of the hotels at Hurghada as well as along the north coast of the country.

**Stonechat** *Saxicola torquata* A common winter visitor of the northern part of the country especially along the Mediterranean and Delta area. The earliest autumn record was on 2nd September 1993 in the Delta near Tanis. A count of 46 was found around Wadi el Natrun with at least 35 around Lake Manzala in 1990. There are few records along the Red Sea with birds found in hotel gardens at both the Grand and the Arabia at Hurghada, with fewer along the coast of the South Sinai.

**Siberian Stonechat** *Saxicola torquata maura* A single bird of this eastern form was found at Sharm el Sheikh from 25th to 27th October 1995.

**Pied Stonechat** *Saxicola caprata* There are 3 records of this species all from southern Israel with an overwintering bird in 1979, one from Negev on 28th October 1994 and another from Eilat from 5th to 7th November 1996. There was also a record from Cyprus in 1997.

**Isabelline Wheatear** *Oenanthe isabellina* This is a common passage bird with smaller numbers actually wintering. It has been commonly seen on Nile cruises with sites such as Giza, Saqqara and Meydum Pyramids having records as well as Hathor, Beni Hassan and El Til. Regular records come from Abu Simbel in the winter months and singing birds have been heard at Dahshur in 1997. In Sinai there is a large wintering population on the Egyptian border at Nizzana where breeding also takes place often in empty fat sand rat burrows

**Wheatear** *Oenanthe oenanthe* A very common migrant with passage covering the country. Even eastern areas hold this species in spring and whenever other species were found it was not long before this bird was found as well. Concentrations of birds have been found at Wadi el Natrun, Faiyum and Lake Manzala with a late record of 26th May recorded at El Malaha in 1990.

**Pied Wheatear** *Oenanthe pleschanka* This is a rare passage and wintering species with an increase of records in the 1990s due to better coverage. The main wintering records come from Abu Simbel where there were two females during 1994/95 and 2 more there during 1995/96. Away from here, there were 3 birds found in Wadi El Arish on 9th October 1994 and 2 immatures at Zaraniq on 17th October 1994. Records again are increasing with the general holiday-maker/birder with Hurghada having 2 females on 25th April and also a single on 7th May 1996 and also another single at Taba on 4th October 1996. Nile cruises have picked up this species at Beni Hassan, El Amarna, and Hathor where the last record was on 1st October 1992.

**Cyprus Pied Wheatear** *Oenanthe cypriaca* This bird has increasing records in the 1990s with 2 at Abu Simbel on 23rd January and again on 24th October 1992, one at Zaraniq on 17th October and two at 10th Ramadan sewage ponds on 23rd October 1994. Five wintering at Abu Simbel during 1994/95 and 2 birds wintered during 1995/96. There was also a single male seen near Hamam Faraun, South

White crowned Black and Mourning Wheatears

Sinai on 25th March 1996. A small passage has also been seen from Port Said and Zaraniq.

**Black-eared Wheatear** *Oenanthe hispanica melanoleuca* A very common passage bird with small numbers irregularly wintering, especially in South Sinai. There is a small breeding population on the Egyptian border with Nizzana, Israel and birds have also been found in the Rafa area. This bird was regularly seen by Nile cruises with Meydum, Beni Hassan, El Til and west bank, Luxor especially. One wintering record came from Abu Simbel during 1994/95.

**Desert Wheatear** *Oenanthe deserti* A breeding bird of mainly the northern desert areas with a large influx of migrants wintering in other areas. This bird was commonly found on Nile cruises around most of the ancient sites visited, with the road to Faiyum also turning up birds. Wintering birds were found at Abu Simbel where they were the commonest wheatear in January 1997. Other records come from Wadi el Natrun, Zaraniq and many areas in South Sinai where they also breed in the mountains.

**Finsch's Wheatear** *Oenanthe finschii* This is a fairly common winter visitor to the north west Negev with around 300 birds and these generally can move over to the Egyptian side of the border. The only recent records come from Wadi El Arish where a female was present on 28th December 1994 and by the rubbish tip again on 16th January 1995

**Red rumped Wheatear** *Oenanthe moesta* A local breeding bird of the western Mediterranean coast. Anywhere west of Alexandria seems a likely place to find this bird but the only recent record comes from the El Omayed protected area near Hamma, El Alamein where a pair was found up to 1998. The mass coastal development here may have a negative effect on this population. There are a few records away from this coast with a small population breeding on the Negev/North Sinai coast and even records inland at Giza pyramids with an old record, and 2 possibles were seen here in 1990.

**Red Tailed Wheatear** *Oenanthe xanthoprymna* This is a rare passage bird but a common winter visitor to Gebel Elba where several hundred are found. Birds have wintered over the years especially at Saqqara where there was a single bird in the winter of 1991/92 and again in 1996/97. The bird was found on Nile cruises with records again from Saqqara and Beni Hassan. On Sinai there was a single bird 40 kilometres south of Taba on 29th December 1993 and one at St. Catherine's on 14th August 1994, with a wintering bird at Sharm el Sheikh during 1997/98.

**Eastern Pied Wheatear** *Oenanthe picata* A single record comes from near Eilat of a dark morph variety on 4th February 1986.

**Mourning Wheatear** *Oenanthe lugens* This is a resident breeding bird of mainly the eastern part of the country especially in the mountains and wadis of the Eastern Desert and Sinai Peninsula. It has been commonly found on Nile cruises at such sites as Beni Hassan, El Amarna, and the Valley of the Kings where there is a local breeding population which is not found in many other sites this side of the river. One of the main sites, for this bird, from Cairo is at Wadi Hagul just off the Suez road. In Sinai the bird is common around Sharm el Sheikh.

**Hooded Wheatear** *Oenanthe monacha* This is another bird breeding mainly east of the Nile with a small population around the Valley of the Kings where it was often found by Nile cruises. Other sites where it was found included Beni Hassan, El Amarna and Abydos. It has been found often by birders along the Red Sea with Bir Beida a favourite site as well as the Marsa Alam to Edfu road for this species while the east coast of Sinai is also popular with some records also from the mountain wadis.

**White crowned Black Wheatear** *Oenanthe leucopyga* This is a favourite bird of the Nile cruises and is a resident breeding bird of mainly the eastern side of the country with many records from around Lake Nasser and into the Western Desert oases. The Nile cruises commonly see this bird around the rocks on their boat trip to Philae and even sitting on top of the High Dam. At Abu Simbel it can be seen at the airport, by the temples and around the hotels. One pair in 1997 were nesting in a toilet overlooking the lake here!

**Black Wheatear** *Oenanthe leucura* Two birds were seen at Quiser on 6th November 1982 the same year when a single turned up at Eilat from 11th to 15th December. The Eilat bird was photographed and after a great debate was the first accepted record for the Middle East. The Quiser birds are left in limbo as others have already called them 'funny eastern pied' or white crowned black wheatears. Was the bird at Eilat one of the Quiser birds?

**Rock Thrush** *Monticola saxatilis* This is a more common spring migrant than an autumn one with records from around the country especially along the Nile where it was seen on Nile cruises. Several sites include the west bank of Luxor, Karnak and Hathor. Records also come from Abu Simbel, Abu Sir and hotels at Hurghada. On the Mediterranean coast the hotel at Al Alamein seems to be a favourite site for them with records from here up to 28th February 1997.

**Blue Rock Thrush** *Monticola solitarius* This is another of the birds which Nile cruises often see while visiting the ancient monuments as it is a fairly common winter visitor to, especially, the Nile Valley. Most sites hold records of this bird which is always seen flying away! Favourite sites for the bird include Abydos, Hathor, the Valley of the Kings and Philae. Occasionally the bird has bred in the country with records from Giza, Salum and Cairo. A few thousand pairs breed in the north of Israel.

**Ring Ouzel** *Turdus torquatus* An uncommon migrant with sometimes a small wintering population. Records from the Mediterranean often come from Al Alamein. Wintering birds have been found mainly on Sinai with several birds found around St. Catherine's Monastery and Wadi Feiran.

**Blackbird** *Turdus merula* A fairly newly-arrived breeding resident from across the border with Israel in the 1970s. It has now spread along the North Sinai coast, into the Delta and down into Cairo and along the Nile to Beni Suef. New breeding records came from the Faiyum and Wadi el Natrun in 1990 and 10 pairs were breeding at Siwa Oasis in the Western Desert in 1994 and seen in other Oases in 1998. Birds have been seen on Nile cruises from Memphis and in Cairo itself.

**Eye browed Thrush** *Turdus obscurus* The first record for Israel and Eilat was on 17th October 1996.

**Black throated Thrush** *Turdus ruficollis* A rare winter visitor with a single record of a male from El Arish on 3rd January 1982. Small numbers have been found at Eilat in the 1980s.

**Fieldfare** *Turdus pilaris* This is an irregular winter visitor with birds mainly found in the northern region of the country. There were 61 birds at Wadi el Natrun on 30th December 1989 and on 15th January several of these were dead, having been shot by foreign gunmen! The hotel at Sidi Abd el Rahman held several birds on 24th/25th November 1990. Nile cruises have records as far south as Luxor, and in Sinai birds have been found in the mountains around St. Catherine's and Wadi Feiran with one at Sharm el Sheikh on 11th January 1995. Several thousand can be found in the north of Israel in most winters.

**Song Thrush** *Turdus philomelos* This is a regular winter visitor with larger numbers in certain years with records as far south as Abu Simbel where there was one on 1st March 1998. There are several areas around Cairo where it can be found including the Gezira Sports Centre, the Agricultural Museum and the Giza Zoo. Nile cruises have found the bird as far south as Aswan on Kitchener's Island but more have come from the Winter Palace gardens in Luxor. The hotels along the Red Sea are starting to pick up a number of birds while the north coast of Sinai seems the more likely place to find them.

**Redwing** *Turdus iliacus* This can be a rare winter visitor with no records in some years. Most records do come from the north Mediterranean coast where several birds were found at Al Alamein on 24th/25th November 1990 and into the Delta with a single bird found at Gebel Asfar on 15th January 1992.

**Mistle Thrush** *Turdus viscivorus* An irregular winter visitor to mainly the Sinai with records from St. Catherine's and Wadi Feiran and also the Delta where a single bird was found at Lake Maryut on 18th January 1990. There was a single bird at Al Alamein in the western Mediterranean on 24th November 1990.

**Cetti's Warbler** *Cettia cetti* An uncommon passage migrant with few records. A number of records have come from Sidi Abd el Rahman [Al Alamein?], the western coast of the Mediterranean with the recent ones being on 20th March 1994 and again here in October 1994. A single record comes from Lake Nasser at the beginning of September 1996. The bird is rare at Eilat but breeds in the far north of Israel.

**Fan tailed Warbler** *Cisticola juncidis* This is a classic bird of the cultivated areas of the Nile Valley and Delta including the Faiyum with many records from Nile cruises. The cruises often see this species when having to walk to the ancient monuments from the boat or on a coach visit, they wander down to the farmland. Hotels which have them close by include the Movenpick on Crocodile Island at Luxor but no birds are found at the Red Sea coast hotels.

**Graceful Warbler** *Prinia gracilis* One of the first birds to waken the visitors on a Nile cruise with this common breeding bird using the smallest amount of vegetation to nest in, often by the boats. They are found along the Nile and into the Delta with expansions into the Faiyum and spreading out around Lake Nasser. This bird is also found breeding out on the salt marshes of the Delta and east onto the Israeli border.

**Scrub Warbler** *Scotocerca inquieta* This is a resident breeding bird of the North Eastern Desert and Sinai. Some of the sites near Cairo are at Wadi Digla near Maadi and along the Suez road in Wadi Hagul and in the Gebel el Galala area. The Sinai population is found in the mountains of the south and down the wadis especially running to the east. Nile cruises have given some records from Abu Simbel but no other records come from this area.

**Pallas's Grasshopper Warbler** *Locustella certhiola* A single record comes from Eilat on 25th February 1983.

**Grasshopper Warbler** *Locustella naevia* This bird is nearly as rare as the previous species with only a handful of records coming from Suez and Kom Ombo. A single bird was seen in the open at the Movenpick Hotel, 20 kilometres north of Hurghada on 21st February 1997, running about the short scrub with 3 Sardinian warblers.

**River Warbler** *Locustella fluviatilis* This species is one that can be easily missed if it adopts its normal habit of skulking in vegetation away from the preying eye. When they are forced to feed where they dropped on migration like the one feeding on the beach at Hurghada on 26th April 1990, the records will appear. The use of mist nets for ringing has shown some sort of movement in Bahig, west of Alexandria on the Mediterranean coast and in Bahariyya Oasis, Western Desert. A single bird was found in the Eastern Desert during March 1993. Another record comes from Abu Simbel on 11th September 1996.

**Savi's Warbler** *Locustella luscinioides* When investigating reed-beds it is either a good view or a song which will identify these birds using such habitat. Several records have come from song in spring or from mist netting at passing sites such as Suez and Bahariyya Oasis, Western Desert where up to 36 birds were caught in 1983. There was a single bird seen at the Sheraton Hotel gardens, Hurghada on 5th May 1991 and another at Al Alamein on the Mediterranean coast on 20th March 1994. Up to 5 were found from Gebel Asfar to Aswan between 19th March and 30th March 1995. Small numbers have been seen by Nile cruises, mainly found in song with records from Dendera, Kom Ombo and Aswan.

**Aquatic Warbler** *Acrocephalus paludicola* This is a very rare bird in the country with one at Hurghada on 20th October 1992 being only the second record for the country. There are no records in Israel.

**Moustached Warbler** *Acrocephalus melanopogon* This was thought to be a fairly common winter visitor and passage migrant but few recent records exist to prove this. Wintering birds were found on Crocodile Island in 1990 with another winter record 77 kilometres west of Alexandria on 22nd January 1993 and also 6 were found at Al Alamein the following year. Passage has been recorded at Suez and Sharm el Sheikh.

**Sedge Warbler** *Acrocephalus schoenobaenus* This is a common spring and autumn passage bird with a small number remaining to winter in the south of the country. Passage in the spring can still be moving through in May with records from Sheraton Hotel, Hurghada and Abu Simbel at this time, but 13 at Wadi el Natrun on 17th May, with 4 singing being exceptional. A wintering bird was found at Abu Simbel on 23rd January 1997.

**Paddyfield Warbler** *Acrocephalus agricola* Several records have come from Eilat, all through ringing and even a bird has been re-trapped here which may have suggested that it wintered close by.

**Blyth's Reed Warbler** *Acrocephalus dumetorum* A few birds have been caught and ringed at Eilat both in the autumn and in the spring. A record from St. Catherine's from 1970 has not been accepted for Egypt.

**Marsh Warbler** *Acrocephalus palustris* This is an uncommon passage migrant with identification a problem with reed warbler especially in its habitat. Singing birds have been found on Nile cruises especially at Abu Simbel where up to 3 males were singing on 3rd March 1985 and again on 31st March 1989. Another bird was here on 9th April 1997. Hurghada has turned up a number of birds in the spring, and passage has been observed at Suez.

**Reed Warbler** *Acrocephalus scirpaceus* It was only in 1983 when the first birds were found breeding in the Delta at Lake Manzala. Many singing birds were found in 1990 even at Lake Qarun and Wadi el Natrun but no proof of breeding was established. Several thousand pairs breed in Israel. Wintering birds were also found in 1990 near Esna, Aswan and Abu Simbel in 1994. It is a common passage bird with many records coming from Nile cruises. In Sinai it is also a common migrant in the north and fewer records come from the south.

**Clamorous Reed Warbler** *Acrocephalus stentoreus* A very common breeding bird and resident of the Nile Valley and Delta. There are even populations out in the Western Desert at Lake Qarun, Wadi el Rayan and Wadi el Natrun with some records from southern oases. There is also a small population along the Suez Canal area but few records from Sinai with 2 males on the 13th February 1998 at Sharm El Sheikh. It was commonly seen on Nile cruises especially from the boat whenever anchored or as the boat drifted by islands of reeds.

**Great Reed Warbler** *Acrocephalus arundinaceus* A fairly common spring migrant but fewer records are found in the autumn due to the similarities of this bird with clamorous reed warbler. Singing birds have been found in spring with records from Nile cruises at Abu Simbel and north of Luxor. There was a singing bird at Lake Qarun on 30th March 1997. Birds have also been caught in mist nets at Suez and Shakshuk, Lake Qarun.

**Basra Reed Warbler** *Acrocephalus griseldis* There are recent reports of this species turning up along the Red Sea coast with 2 birds at Hurghada in May 1994 and one in the mangroves at Bir Ouei 46 kilometres south of Safaga on 8th May 1996. A possible bird was seen at Sharm El Sheikh golf course on 21th February 1998. These May records match the 3 records at Eilat where the last one was on 24th May 1993. Others have been caught in April and June away from Eilat and one in September at Eilat

**Eastern Great Reed Warbler** *Acrocephalus orientalis* Two records of this species come from Eilat from 28th February to 13th April 1988 and on 2nd May 1990.

**Thick billed Warbler** *Acrocephalus aedon* There is only one record of this species with a single bird being found at St. Catherine's Monastery in Sinai on 20th November 1991. There are no records for Israel and this makes it only the 3rd record for the Western Palearctic with the 2 others coming from as far as Shetland, UK.

**Olivaceous Warbler** *Hippolais pallida* This is a common breeding bird of the Nile Valley and Delta with populations also found around Faiyum, Wadi el Natrun, other western oases, the Suez Canal and the north coast of Sinai. New breeding locations are along the Red Sea with mangroves being used south of Abu Ghusun in 1990. It is also a common migrant with a small population wintering. It was commonly seen by Nile cruises with birds often found in the farmland trees as the guests walked to monuments .

**Booted Warbler** *Hippolais caligata* Several of this species have been caught in mist nets at Eilat both in the spring and in the autumn. This species may well be found in the hotel grounds at Hurghada in the future.

**Booted (Syke's)Warbler** *Hippolais caligata rama* One of the rama race wintered in Eilat from 3rd December 1993 to 7th April 1994.

**Upcher's Warbler** *Hippolais languida* A rare migrant in Egypt but it is a breeding bird of the north of Israel. Birds have been found near Ras el Naqab, South Sinai and near Qusaima, North Sinai in April 1992. A single bird found at Sharm el Sheikh on 11th January 1995 was only the eighth record for the country but the first winter record.

**Olive tree warbler** *Hippolais olivetorum* This is a scarce passage migrant with many more records in the autumn than the spring. It has been found on odd occasions on Nile cruises with records coming from El Til, Aswan and the High Dam. Several have been located in Wadi Sahl, South Sinai and up to 13 seen in Wadi Merah south of Eilat in mid-July showing a very early migration. 1997 saw a strong passage especially along the Red Sea with one at Hurghada on 2nd May and 5-6 at Ain Sukhna on 10th May and one in North Sinai at Ain el Gedeirat. The bird breeds in the far north of Israel where there was a strong autumn passage as early as 25th July 1997 near Eilat.

**Icterine Warbler** *Hippolais icterina* Several birds of this fairly common spring and autumn migrant have been found by Nile cruises over the years. Along the Nile the bird has been found from Aswan, Kom Ombo, West Bank of the Nile and the Winter Palace Hotel at Luxor. The Hurghada hotels are now turning up birds with one even found on the islands on 24th September 1996 and also at Zaraniq, North Sinai where there was a single bird on 6th September 1996.

**Marmora's Warbler** *Sylvia sarda* There are two very old records for this species with one taken along the Western Mediterranean coast at Salum and two birds together from South Sinai. The Mediterranean record is far more convincing as small numbers winter in North Africa. There are no records for Israel.

**Spectacled Warbler** *Sylvia conspicillata* This is probably a breeding bird of the north east Sinai with a large population breeding over the border in Israel with up to 300 pairs nesting near Nizzana. Singing birds have been found here and farther south into the mountains around St. Catherine's and Wadi Feiran and Wadi Hibran. Away from here north there is some passage, with up to 10 having been found north east of Cairo in the Western Desert in November 1985 while singles have been found on Nile cruises with up to two at El Amarna and a single at Karnak. There was also a single in the Eastern Desert at Wadi Hagul on 9th May 1991 and a pair with 2 young were found on 20th March 1995.

**Subalpine Warbler** *Sylvia cantillans* This is a passage migrant with a small number staying to winter in the south of the country. The hotels around Hurghada are picking up many birds both in spring and autumn while the Nile cruises only had a single record from Abu Simbel. There were several birds in spring 1997 with a female by the Red Pyramid at Dahshur. Several more were found with one on Kitchener's Island, Aswan, at Wadi Hagul and more in the sports grounds at Gezira, Cairo.

**Menetries Warbler** *Sylvia mystacea* This is a bird easily identified wrongly as Sardinian warbler and there are only a handful of records from Egypt. Most of these come from Sinai (Nuweiba and Sharm el Sheikh) with another from Moon Mountains on the border with Israel where there has been several records since 1970, mainly at Eilat. Two birds were found at Wadi Hagul, Eastern Desert on 27th November 1993. A single record comes from the Nile cruises with a bird seen clearly at Cataract Gardens, Aswan on 25th March 1987. More records are likely from the expanding hotel grounds of South Sinai and the Red Sea.

**Sardinian Warbler** *Sylvia melanocephala* This is a common migrant and winter visitor to especially the Nile Valley and Delta areas of the country. A small breeding population is found in northern Sinai along the coast and a rare form of this species is now extinct from the Faiyum. Several records come from Nile cruises both in winter and on passage with Abu Simbel being the star place to find it in winter while the gardens around Luxor and Aswan also turned up a number of birds. Many birds now winter around the gardens in Hurghada and other hotel complexes of the Red Sea and Sinai. Most migrants have passed through the country by April.

**Cyprus Warbler** *Sylvia melanothorax* In 1981 a group of ten birds were found in South Sinai; now only single records exist in the 1990s. The birds favour acacia trees and with increased pressure on these trees by so-called 'safari' groups coming out of the tourist centres and felling the trees for fire-wood, fewer sites may now exist for this bird. There was a single bird found at Wadi Zalaga on 12th January 1997. Away from Sinai single birds have been/were found down the Red Sea coast as far as Gebel Elba with a recent record of a male north of Zafarana on 8th March 1996. Up to 300 birds winter in the southern part of Israel with two at Sharm el Sheikh in 1994/95.

**Ruppell's Warbler** *Sylvia rueppellis* A common migrant especially along the Nile with several records from Nile cruises especially around Aswan and Abu Simbel in the spring. It can even be found in the centre of Cairo where birds were seen in the gardens of Garden City in 1997 and in the Gezira Sports Centre. It is commonly found in the hotel complexes of Hurghada and other Red Sea resorts with most birds moving through by April.

**Desert Warbler** *Sylvia nana* This bird has declined in recent years as a winter visitor to South Sinai due to overgrazing by Bedouin goats. There were only 4 found in South Sinai on 25th November 1994 and 3 found in the north at Wadi Gerafi on 21st December 1994 and again on 17th January 1995. Another 3 were found wintering at Gilf Kabir in 1996/97 and another at Gebel Uweinat which are the first wintering sites in the southern Western Desert. There are several old records from the Red Sea coast and Eastern Desert down to Gebel Elba with a single bird found near Marsa Sagra in December 1995. It is also a wintering bird of Sudan. Up to 400 birds have wintered in southern Israel but even now these are threatened by increasing agricultural development encroaching on the desert.

**Arabian Warbler** *Sylvia leucomelaena* Gebel Elba has a small population of this species with birds located in 1997. Small numbers have also been seen in eastern Sinai with birds moving over from the small Israeli breeding population of around 120 pairs in the Arava Valley.

**Orphean Warbler** *Sylvia hortensis* An uncommon passage bird with many more records from the spring, especially from the hotel complexes along the Red Sea. A single bird was found at St. Paul's Monastery on 1st April and 4 more in the desert between Aswan and Abu Simbel on 7th April 1997. It was a rare bird on the Nile cruises with some early February records as well as the normal spring records from Abu Simbel.

**Barred Warbler** *Sylvia nisoria* A rare passage migrant with increasing records expected from the hotel complexes of the Red Sea, Mediterranean and Sinai. There were two at Hurghada on 26th April 1990 and one at Suez on 3rd May 1990. A single bird was found at St. Catherine's on 24th September 1996. The largest gathering of this species was 22 birds from Gebel Elba in April 1985. Sudan, next-door, is a known wintering area.

**Lesser Whitethroat** *Sylvia curruca* This is the commonest warbler after chiffchaff migrating through Egypt. It can be seen so many times it may stop some birders looking through the flocks of warblers! A small population also winters in the country with records from many areas including the Red Sea. It was commonly seen on Nile cruises and ringing information shows a movement out of Western Europe and even Britain where recent recoveries were found farther west than others along the Mediterranean coast.

**Whitethroat** *Sylvia communis* This is another common migrant with some spring records as early as February and autumn ones by the end of July with late records by 22nd November. It was not commonly seen on Nile cruises with many more records coming from the hotel complexes along the Red Sea and Sinai. In 1997 it was found from 9th April onwards with a noticeable decline in lesser whitethroats in the centre of Cairo.

**Garden Warbler** *Sylvia borin* Another common migrant to the country with records even from the Western Desert. The spring records are more common than the autumn with increasing finds in the hotel complexes of the Red Sea and Sinai. Birds were found at the Sheraton Hotel, Hurghada as late as May while a single bird was at Faiyum on 15th May 1990.

**Blackcap** *Sylvia articapilla* Again, like the garden warbler, there are more spring records than the autumn for this widespread migrant. There are few winter records for this species but two birds were in the Agricultural Museum's gardens, Cairo on 14th February 1997 and were commonly moving through the city by the end of March. In South Sinai there was a late movement of 10 males and 9 females on 5th May at Ain Musa.

**Green Warbler** *Phylloscopus nitidus* A single bird ringed at Eilat on 27th October 1987.

**Yellow browed Warbler** *Phylloscopus inornatus* This is an under-recorded species of especially the Sinai, with up to 15 found at Eilat most autumns and smaller numbers in the spring. One was found in the Taba Hilton Hotel Gardens, South Sinai on 20th December 1994, this being the first wintering record.

**Hume's yellow browed Warbler** *Phylloscopus humei* Another interesting species with a single bird found at Sharm el Sheikh from 17th to 18th October 1992 with recent wintering records from Eilat where the last one was found during February and March 1993.

**Dusky Warbler** *Phylloscopus fuscatus* The first record for Egypt occurred on 12th and 13th October 1988 at Naama Bay, Sharm el Sheikh. The bird was so tame that it was easily recorded on a video recorder. There is only a single record from Eilat of this species on 25th March 1989 - the same bird returning north!

**Bonelli's Warbler** *Phylloscopus bonelli* This is a common migrant passing through the country on a broad front. It was commonly seen in the spring on Nile cruises especially around Luxor, Aswan and Abu Simbel and many records come from the hotels around Hurghada where up to 20+ can be found in early April. Smaller numbers have been found at Wadi el Ryan, Wadi el Natrun, Faiyum in the Western Desert and Wadi Hagul, Port Said and Suez on the eastern side, with Sharm el Sheikh offering birds in the Sinai.

**Wood Warbler** *Phylloscopus sibilatrix* An uncommon migrant with increasing records from the hotels at Hurghada especially in April when it was also found on Gezira Island, Cairo along with other warblers. Other records come from Faiyum, Wadi el Natrun and Lake Manzala. Nile cruises managed to find this bird at Beni Hassan and Aswan.

**Mountain Chiffchaff** *Phyllosocopus sindianus* A single record came from the Nile cruises on 22nd March 1986 when it was spotted in the gardens of the Winter Palace Hotel, Luxor. The only Eilat record also comes from March when a single bird was found on 5th March 1983.

**Chiffchaff** *Phylloscopus collybita* This is the commonest of all the warblers passing and wintering in Egypt. From the centre of Cairo to an isolated acacia out in the Eastern Desert you cannot miss this bird. It was commonly seen by passengers on the Nile cruises with the classic saying, 'What is that little brown bird hopping about in that tree?'

**Willow Warbler** *Phylloscopus trochilus* A very common migrant often overlooked due to so many chiffchaffs which is a similar bird. There are very few winter records even though they are supposed to winter in Sudan. An estimated 1,000 birds were present in the Western Desert oases of Bahariya in August 1983. There were far fewer records of this species on Nile cruises and the only increase of records come from hotels around Hurghada.

**Goldcrest** *Regulus regulus* There are few records for this species with most of them coming from the Mediterranean coast with some very old records from Giza. There were 4 birds found at Montaza Gardens, Alexandria in 1992 and 3 there on 23rd January 1993 with up to 5 birds wintering in the area. There is a small wintering population in Israel but few records from Eilat.

**Firecrest** *Regulus ignicapillus* There is an old record from 5th March 1973 from Bahig, west of Alexandria, on the Mediterranean.

**Spotted Flycatcher** *Muscicapa striata* A fairly common migrant with a scattering of winter records. Nile cruises have found this bird both in the spring and the autumn with a single record from Giza Zoo in February. Migration leaves this bird in often desert conditions with a single bird found on a 'koppie' out in the Western Desert, 10 kilometres north of Abu Simbel on 7th April 1997. No other birds were with it. The hotel complexes of Hurghada are turning up more and more birds while a very small breeding population is found along from El Arish, North Sinai to the Israeli border where a larger breeding population exists of a few thousand pairs, especially in northern Israel.

**Red breasted Flycatcher** *Ficedula parva* An uncommon passage migrant with a scattering of records around the country. Larger numbers are seen if ringing and mist nets are used, with up to 21 caught in the Western Desert at Bahariyya in 1983. Recent records include one at Abu Simbel in October 1992, 3 found in the grounds of the Al Alamein Hotel at Sidi Abed el Rhman, Marsa Matruh, on the Mediterranean between 30th September and 4th October. A single bird was found in an orchard south of El Arish, North Sinai on 7th October and another one at 10th Ramadan sewage works on 19th October 1994. Three were found at St. Catherine's from 23rd to 24th September 1996. There were at least 4 in gardens at Marsa Matruh on 27th September 1997. The rarer spring records include one at Zaraniq on 28th April 1996 and another at Ain Sukhna on 11th May 1997. One wintering at Sharm El Sheikh found during February 1998.

**Semi-Collared Flycatcher** *Ficedula semitorquata* This is the rarer of the three 'pied' flycatchers with birds mainly seen in the spring, especially from Gezira Sports Centre and along the Mediterranean coast. There was a possible female of this species seen on 28th April 1990 and another at the Jasmine Hotel at Hurghada on 2nd May 1997. Up to 50 have been seen at Eilat again in the spring.

**Collared Flycatcher** *Ficedula albicollis* A commonly-found bird of the spring but far fewer records in the autumn. Surprisingly this bird was not found on Nile cruises with most records coming from Gezira Island, Cairo and Hurghada, with a few records from the Western Desert oases

**Pied Flycatcher** *Muscicapa hypoleuca* Like the collared flycatcher there are few autumn records but many more spring ones from similar localities. It was seen side by side with collared in the Sports Centre at Gezira in the spring of 1997. Records have also come from the Western Desert oases, Marsa Matruh, CrocodileIsland, Luxor and Abu Simbel. Two late records come from Lake Manzala on 8th May 1990 and Suez on 13th May 1990. The latest autumn record was 9th November.

**Bearded Tit** *Panurus biarmicus* The only record for this species comes from Lake Manzala on 31st December 1987. Five birds were seen with many more heard. It is very rare in Israel with all records from the north of the country and it is classed as an irregular winter visitor to Cyprus.

**Arabian Babbler** *Turdoides squamiceps* This is a species restricted to the north eastern half of the Sinai with small groups located as far south as Wadi Taba with 8 found near Ras El Naqab, southern Sinai on 8th March 1995. These were the first here since 1982. Singles have also been seen at Nuweiba in the late 1980s. There is a resident population in the Eilat area with occasional birds being forced over the border by the fall-outs in their social parties.

Cyprus Pied Wheatear

Abyssinian Roller

White crowned Black Wheatear

Lanner Falcon

Sewage, Storks and Sina

Nabq

Ras Mohammid

St Catherine's Monastery

Chukar

Little green Bee-eater

Namaqua Dove

Red rumped Swallow

Steppe Eagle

Grasses in the Sinai

Cyprus Warbler

Blackstart

Rose-coloured Starling

Citrine Wagtail

Common Sandpiper

Red Storks!

**Fulvous Babbler** *Turdoides fulvus* There was a former colony around Toshka along the Nile above the High Dam which was destroyed after its completion and flooded their territories. This left the only breeding area as Gebel Elba with the hope of future colonisation of the birds from the Sudanese side of the border along Lake Nasser as vegetation develops.

**Great Tit** *Parus major* A rare breeding bird of the North Sinai coast as far west as El Arish with young found in Wadi el Arish in June 1990 and also at Rafa where family parties were also seen in 1990. There are several thousand pairs breeding in Israel with movement allowing one bird to visit Eilat in 1983.

**Penduline Tit** *Remiz pendulinus* With the increase in birders, more sightings than ever before have been recorded in recent years. The first record for the country appeared only in 1976 with most birds found at Suez where there were up to 10 birds moving through in 1990. There were 6 here on 20th March 1992 with other birds coming from Lake Manzala on 5th March 1990 and 2 on 19th February 1997. Wintering birds have been found in the Sinai with 10 around Wadi Feiran in 1988/89 and two at Sharm el Sheikh on 16th November 1995.

**Nile Valley Sunbird** *Anthreptes platurus* This is yet another special bird to be found in Egypt. It can be found from Cairo to Aswan where non-native trees and shrubs have added to their feeding range and this is probably the reason for its movement north to Cairo as a breeding bird. The sports area on Gezira Island, Giza Zoo, the Agricultural Museum and Garden City are some of the many sites where you may find this species in Cairo. The gardens of many hotels from Luxor to Aswan harbour this bird. Birds are now also found in the Faiyum.

**Shining Sunbird** *Nectarinia habessinica* Gebel Elba is the only breeding site in Egypt for this species. With restricted access to the area there seems little chance of most birders being able to see this species in this country at the present time. Ten were seen on 31st March 1997 by one group entering for the first time. There is some movement locally in the area due to the flowering of shrubs and trees taking place at different times of the year. Display was seen in December 1997.

**Orange tufted Sunbird** *Nectarinia osea* A small population breeds along the north coast of the Sinai with increasing numbers moving into central and eastern parts of the Sinai. A pair were found at Wadi Kidd which is only 30 kilometres north of Sharm El Sheikh on the 26th February 1998. It is a common migrant in Israel with 100,000+ pairs breeding. Recent records have included 4 flying west at Zaraniq on 18th September 1996 which is the farthest west the species has ever been seen and a new breeding site for the bird was at St. Catherine's Monastery in 1996 with a male on 18th May 1997.

**Golden Oriole** *Oriolus oriolus* A common migrant both in the autumn as well as the spring with one exceptional record from El Hauwariya to the south west of Alexandria where thousands of birds were seen on 27th April 1982. Generally they are seen singly or in groups of up to five with their call giving them away when perched in trees. The island of Gezira Abu Minqar off Hurghada held 13 birds on 27th April 1990 while Nile cruises bumped into the birds especially at Aswan on Kitchener's Island where up to 14 were seen on 16th September 1993. The hotels around Hurghada are turning up a number of birds each year.

**Rosy patched Shrike** *Rhodophoneus cruentus* One of the many special birds from Gebel Elba with up to 15 seen on 31st March 1997. Is not found anywhere else in the Western Palearctic.

**Isabelline Shrike** *Lanius isabellinus* There were few records in the past but since 1995 the number has increased dramatically. A male was found wintering at Abu Simbel during 1994/95. Another male was found 25 kilometres south of Aswan on 25th March 1995 followed by another 50 kilometres west of Quiser on 13th April 1995. There was a single bird at Wadi Kid, Nabaq, South Sinai on 29th March 1996 and another single bird at Bir el Ambagi on 9th May 1996 near Quiser. There was also another wintering bird at Abu Simbel in 1996/97 and 97/98 and Gebel Elba in December 1997 with up to 8 birds seen each year at Eilat.

**Red backed Shrike** *Lanius cristatus* A much commoner autumn bird than spring with large numbers sometimes seen in the Sinai. The oasis at Bahariyya, Western Desert had 170-200 birds on 3rd September 1983 while 90 birds were found in 500 metres of acacia in the Gulf of Suez. There were few records on Nile cruises with early spring records from February on Kitchener's Island, Aswan and Kom Ombo. Later spring records come from Hurghada where 5 males were found on 26th April 1990. Small numbers are known to winter in the Western Desert. Over the years there have been many ringing returns of birds migrating south, especially from Sweden.

**Lesser grey Shrike** *Lanius schach* Like the red backed shrike there are far more autumn records than spring ones except along the Red Sea where the birds seem to be common in both seasons. Small numbers were seen on Nile cruises at Lake Qarun, Assiut, Hathor with 2 on 22nd September 1993 and at Luxor. A recent spring record came from west of Alexandria on 28th February 1997 and another from Al Alamein on the same day. Red Sea records include one at Safaga on 22nd and 23rd March 1995 and 4 were on the Port Said-El Cap road, Ismailiya on 30th April 1990. A total of 20 were counted at Sharm el Sheikh on 21st August 1986.

**Great grey shrike** *Lanius excubitor* A breeding bird of the Nile Valley and Delta with smaller numbers along the Red Sea and Sinai. A pair at Lake Qarun were watched mobbing a little owl for around 10 minutes while the tide had turned on a juvenile which was being mobbed by 6 olivaceous warblers at Beni Hassan. They have even been found for sale as food in the bird markets in the Delta area; not much fun for the 'butcher bird'. Commonly found on Nile cruises especially as the groups walked to temples and pyramids at Meydum, Abydos, Hathor and the west bank of the Nile at Luxor.

**Steppe Great grey Shrike** *Lanius excubitor pallidrostris* A single bird of this sub species was found in Wadi Gerafi, North Sinai on 21st December 1994. One was at Eilat on 20th November 1987 but is thought to be an overlooked species for the country.

**Woodchat Shrike** *Lanius senator* A very common spring migrant with far fewer records in the autumn. Nile cruises often found this bird at Aswan especially on Kitchener's and Elephantine Islands, with other records from Edfu and Hathor. Birds have been found in good numbers along the Red Sea and Sinai with Hurghada hotels turning up many. There are a few thousand pairs breeding in Israel with birds found right down to the adjoining Gaza Strip.

**Masked Shrike** *Lanius nubicus* Compared with the woodchat shrike there is more of an eastern migration through the country with fewer records north of Qena. It was commonly found on spring cruises on the Nile with records from Luxor, Aswan and Abu Simbel. Crocodile Island, Luxor was a favourite site for this bird where even display and territorial behaviour have been seen. The Red Sea hotel gardens are a magnet for birds and a late June record of a male at Rafa may have been a non-breeding bird from the Israeli population of a few thousand pairs.

**Jackdaw** *Corvus monedula* Old records for the country suggest a mistake was made with house crows but with 20,000 wintering birds next-door in Israel new records may be around the corner, especially near Rafa.

**House Crow** *Corvus splendens* A common breeding bird of the Suez and Port Taufiq area with an estimated population of up to 4,000 birds which breed in the park trees and even with the cattle egrets at Suez. Around 80 birds were seen at Ataqa, 20 kilometres from Suez on 31st March 1997. Small numbers have been seen away from this area from Port Said to the north and down to El Tor in the South Sinai. Up to 32 birds are now found at Eilat.

**Rook** *Corvus frugilegus* Small numbers have been recorded in Egypt with most records coming from the Red Sea coast in the 1970s. There has been a dramatic decline of this species in Israel due to chemicals used in agriculture and only now are numbers increasing again in the winter population which may add new records to Egypt with small numbers found in north west Negev.

**Hooded Crow** *Corvus corone* A common breeding bird of the Nile Valley and Delta with birds moving along the north coast of Sinai. Birds can be very tame in some of the hotel gardens where they walk around close to your feet. Their nests can be used by both great spotted cuckoo in the upper Nile and hobby along the north Sinai coast, and kestrels all over the crow's range. A recent record came from Mangrove Bay, south of Quiser on the Red Sea, of a single bird on 1st May 1997.

**Brown necked Raven** *Corvus ruficollis* A well sought-after species in the country with many records coming from around the ancient sites and even nests have been found on the White Pyramid at Dahshur, on the Great Hypostyle Hall at Karnak and on the temple of Hatshepsut. Nesting material has been seen carried as early as January 3rd and large numbers can be found around rubbish dumps, especially at Abu Simbel where up to 120 have been counted.

**Raven** *Corvix corvix* This is the rarest of the three ravens found in Egypt with just a handful of pairs breeding along the Western Mediterranean coast. Few records come from this area with a single bird seen along the Marsah Matruh-Salum road on 16th March 1994 and a party of 8 birds in 1997. Migrants have been seen at Eilat so odd records may well come from Sinai at some point.

**Fan tailed Raven** Corvus rhipidurus This species is found in only two areas of Egypt with the southern range down around Gebel Elba and the other site being in South Sinai. The Sinai birds are easily seen on a visit to St. Catherine's but away from here there were 2 at Sharm el Sheikh on 7th May 1996. They were found on the outskirts of Dahab on 4th February 1997 where there were up to 16 feeding on rubbish and 10 outside Abu Rudeis on 5th February 1997. There were 26 seen at Gebel Elba on 31st March 1997.

**Tristram's Grackle** *Onychognathus tristramii* A bird restricted to the South Sinai mountains with St. Catherine's the main site for this species. Many other local wadis hold birds with records from Wadi Feiran, Kid, Nasb and Zaghra. There is a breeding population at Eilat spreading north with up to 2,000 pairs but migration here indicates that birds come out of the Sinai mountains, with up to 50 birds visiting Eilat.

**Starling** *Sturnus vulgaris* Most winters this is a common visitor to the Mediterranean coast and inland into the Delta. With a project working in the Delta in 1990 there were several records from this region with 600 in the Lake Manzala area up to 12th January. A large gathering of 2,510 were found between Bilgas and Baltim on 13th January falling off until 22nd March when the last birds were seen. There were 2 birds at Gebel Asfar, north east of Cairo on 19th March 1995 and 10+ birds at El Hammam, west of Alexandria on 28th February 1997.

**Rose-coloured Starling** *Sturnus roseus* This is a rare migrant to, especially, the Sinai from where several records have come in recent years with birds even seen in Wadi Feiran and at St. Catherine's. There was one on 15th and 23rd August 1995 and another on 22nd September 1996 both at Sharm el Sheikh, South Sinai. It is a regular migrant at Eilat with up to 20 birds in the autumn and fewer in the spring.

**House Sparrow** *Passer domesticus* Seeing this bird in Egypt gives some idea of the mass flocks formerly seen in Europe before chemicals were used. On Nile cruises I have seen thousands flying up to roost by the river but have also seen kestrels, sparrowhawks, sooty falcons and black shouldered kites catching these birds as part of the food chain. Locals feed on them as they used to do in Europe and still there are plenty left. Large roosts also occur in Cairo and many towns now. The normal breeding distribution is around cultivated areas and into villages away from the desert.

**Spanish Sparrow** *Passer hispaniolensis* A very common winter visitor with birds found from the western Mediterranean down as far as Abu Simbel and east along the Red Sea and into Sinai. There were up to 150 nests at Dakhla Oasis in February 1998. There was a male outside a nest at El Alamain on 1st March 1997 and there seems to be some suggestion of breeding birds before this at El Arish. There are tens of thousands of pairs next-door in Israel with many pairs also in Libya. There were a few sporadic records on recent Nile cruises as far south as Edfu on the American trips. There were 300+ on Crocodile Island on 27th January 1997 and several records from Hurghada and its hotels with birds moving through Abu Simbel to 3rd May 1991. There were several birds at Marsa Matruh on 26th September 1997.

**Dead Sea Sparrow** *Passer moabiticus* This is a rare migrant with few recent records but passage through Eilat may indicate that more birds should be present on this side of the border. It now breeds at Eilat.

**Desert Sparrow** *Passer simplex* A rare bird of the south western corner of the country. As so few birders travel this way there are few records from here. It is known to breed in Libya.

**Tree Sparrow** *Passer montanus* A single bird at Zaraniq from 28th April until 1st May was only the 4th record for the country. Two other records occurred in the 1980s along the Red Sea coast. It is an extremely rare bird in Israel.

**Sudan golden Sparrow** *Passer luteus* The last records come from Gebel Elba in the 1980s when up to 29 birds were seen here and old nests of possibly this species were found. There are no recent records of this well-known wanderer.

**Pale rock Sparrow** *Petronia brachydactyla* There have been some 'invasion' years with 1988 outstanding with many thousands moving through Israel, and 600 at Eilat. Few records came from this side of the border even though 70 were at Nizzana. Latest records include one at Sud Rawafah, North Sinai in April 1992 and 2 were found at Bir Beida in Wadi Abu Ziran, Quiser on 13th April 1995.

**Yellow throated Sparrow** *Petronia xanthocollis* A single record comes from Eilat on 10th/11th May 1982.

**Streaked Weaver** *Ploceus manyar* This is an escaped bird from 1971 with breeding confirmed in 1978, with a population developing around the Delta region. Sites visited to find this bird include the Nile Barrage, north of Cairo, Gebel Asfar and Abbassa both north east of Cairo. Birds were seen in the Faiyum in 1984 and several breeding birds were seen in Lake Manzala in 1990.

**Ruppell's Weaver** *Ploceus galbula* A weaver's nest was found on 1st January 1995 in Wadi Seih, South Sinai. No birds were present but given that streaked weaver is a water bird it was not thought to be this species. Ruppell's weaver was the most likely candidate with breeding birds as far as Riyadh in Saudi Arabia - 1,500 kilometres as the weaver flies! The nest was a classic 'dry' weaver being found in an acacia with the entrance hole to the side.

**Avadavat** *Amandava amandava* Another escaped species which has now spread not only in the Delta but south as far as Aswan on the Nile. It has been seen on Nile cruises at such locations as Lake Qarun, Memphis and Jacob's Island, Cairo. Up to 70 were found on Crocodile Island, Luxor in the spring of 1997. There were 6 at Aswan in November 1997. Gebel Asfar is a common place to see this species near Cairo.

**African Silverbill** *Euodice cantans* A possible breeding bird of Gebel Elba with records from April 1985.

**Chaffinch** *Fringilla coelebs* A bird of the winter months with large numbers in some years. In 1990, 160 were found at Wadi el Arish in February 1998, 30 at Baltim in the Delta on 16th January with 5+ seen around Port Said in March. St Catherine's had 25 birds on 2nd January 1995 and Sharm el Sheikh held 5 birds on 7th January 1995 and one in February 1998. A single bird was found at Hurghada on 23rd March 1995. It is a very common winter visitor to Israel.

**Brambling** *Fringalla montifringilla* A much rarer bird than the chaffinch with records mainly from Sinai with a number coming from St. Catherine's. There was a single bird found at Sharm el Sheikh on 27th October 1995. The hotel Al Alamein at Sidi Abd el Rahaman on the western Mediterranean coast has held autumn birds.

**Red fronted Serin** *Serinus pusillus* There are only two records for the country, all before 1970, with one from Bahig, Western Mediterranean and 3 birds from Bir Naba, South Sinai. The bird is rare in southern Israel.

**Serin** *Serinus serinus* At least 20 pairs were breeding at Rafa on 30th March 1996 with young also seen. This is only the second site for the country with another one in the north west part of the Delta at Gianacilis where up to 6 pairs have bred. Birds have been seen well inland as far south as Cairo with most records from North Sinai in winter. There is a large wintering population in Israel with a breeding population established especially in citrus groves up to the border.

**Syrian Serin** *Serinus syriacus* This is a rare winter visitor to especially around St. Catherine's where up to 20 birds have been found. From the 17th to 19th October 1994 a pair were present at Orabi Farm, 25 kilometres west of Cairo and 8-10 birds were found near Ras el Naqb, South Sinai on 11th January 1997. It is a winter visitor to Eilat and a spring migrant.

**Greenfinch** *Carduelis chloris* A local breeding bird of the Delta and North Sinai coast with a larger population wintering. Singing birds are even heard around Saqqara Pyramid and Wadi el Natrun. Two birds were found at the Nile barrage on 31st March 1995 and a male was seen at St. Catherine's Monastery on 4th February 1997.

**Goldfinch** *Carduelis carduelis* This is a common breeding bird of the cultivated areas from the Delta all the way south to Aswan. It was often found by Nile cruises and always seems to be out of place in such a hot country.

**Siskin** *Carduelis spinus* This is a rare winter visitor to the country with records mainly from the Delta with Montaza gardens, Alexandria holding 13 birds through the winter of 1992/93, and Sinai especially from El Arish, St. Catherine's and Wadi Feiran. One was west of Safaga, Red Sea on 30th November 1993 and 2 were at Al Alamein at Sidi Abd el Rahaman on 20th March 1994 and one again here on 28th February 1997.

**Linnet** *Carduelis cannabina* A fairly common winter visitor to the Mediterranean coast and Delta area with records even from Wadi el Natrun and Faiyum. An exceptional spring record comes from Suez on 4th May 1990. Several

Nile Valley Sunbird

flocks were observed in Marsa Matruh on 1st March 1997. Sharm El Sheikh held 40+ birds in February 1998.

**Desert Finch** *Rhodospiza obsoleta* Breeding was confirmed at Wadi El Arish and Sheikh Zayed in 1995 with young seen. Old records suggest that this was just a winter visitor but up to 150 pairs were found breeding in the north west Negev over the border and now the bird even breeds around Eilat. A flock of 10 birds were found at Sharm el Sheikh on 30th December 1994 and were probably wintering in the area. Single birds were again present at Wadi el Arish on 28th March 1996 and 2 were there on 8th May 1997.

**Trumpeter Finch** *Bucanetes githagineus* A special bird wanted by most birders visiting the country and the best sites are in the Valley of the Kings and Queens on the west bank of the Nile at Luxor. Up to 200 birds have been seen at the Kings and 40 at the Queens in recent years with most Nile cruises finding them somewhere along the Nile. Other sites have included Beni Hassan where up to 150 have been seen feeding in the agricultural fields, El Til and Abydos. Its main range seems to be very much eastern with few records in the Western Desert.

**Scarlet Rosefinch** *Carpodacus erythrinus* A rare migrant to the country with small numbers caught at St. Catherine's in the past, and along the Mediterranean. A single juvenile was found at El Arish on 7th October 1994 and there was a single bird at Sharm el Sheikh on 23rd September 1996.

**Sinai Rosefinch** *Carpodacus synoicus* This is a breeding bird of the mountains of South Sinai where several hundred pairs breed. They are found in lower ground during some winters but the place to see most regularly this species is at St. Catherine's Monastery where they breed and winter. Up to 15 were seen here in May 1997 and 4 were found at a wadi near Wadi Sudr, far north and west of their normal range on 12th March 1997.

**Hawfinch** *Coccothraustes coccothraustes* This is an extremely rare winter visitor with few recent records. One was found at Wadi el Arish on 13th January 1990 and one bird was found on 12th November 1996 at Alexandria. There was an 'invasion' year of this species in Israel during the winter of 1989/90.

**Yellowhammer** *Emberiza citrinella* There is only a single record of this species coming from Heliopolis, Cairo on 21st February 1980. There are 4 records from Eilat and one from Nizzana of a bird which normally winters farther north in Israel.

**Cirl Bunting** *Emberiza cirlus* A very old record comes from Damietta next to Lake Manzalla from 1879.

**Pine Bunting** *Emberiza leucocephalos* There is a single record of this species from Eilat on 1st November 1987 along with a hybrid pine bunting/yellowhammer cross in 1984.

**House Bunting** *Emberiza striolata* There are three distinct populations with one in the south west corner of the Western Desert around Gebel Uweinat, another at Gebel Elba and a more likely area for the birders to see this species in South Sinai. The monastery at St. Catherine's is a likely place to see this species and also at Wadi Feiran. Recent records include one at St. Catherine's on 21st March 1992 and at least 6 at Taba on 16th October 1994 which may be a resident site for this bird.

**African (cinnamon breasted) Rock Bunting** *Emberiza tahapisi* A male was photographed on 20th April 1984 at St. Catherine's. This is the only record for Egypt and has not occurred in Israel.

**Cinereous Bunting** *Emberiza cineracea* A rare bird confined to Sinai with one record from Gebel Elba. Most records are in the spring with the latest record being of a female found 10 kilometres east of El Arish on 30th March 1995. Eilat can have up to 5 in the autumn and 20+ in the spring months.

**Ortolan Bunting** *Emberiza hortulana* An uncommon autumn and spring migrant with records coming from a wide area. Recent records include one from Hurghada on 21st March 1995 and one from Wadi el Rayan on 30th March 1997. A very late spring record comes from South Sinai on 5th May 1990 when 3 birds were seen at Ain Musa. A very early record came from Nile cruises with one near Kom Ombo on 23rd February 1976.

**Cretzschmar's Bunting** *Emberiza caesia* This is a common autumn and spring migrant with most records from the Sinai with Zaraniq the best place to find them in especially the autumn when they can be seen side by side with ortolan bunting as a comparison. There were 3 birds at Suez on 21th February 1998.

**Rustic Bunting** *Emberiza rustica* There is an old record from Bahig, west of Alexandria on the Mediterranean on 22nd October 1972. This bird is commoner in the autumn at Eilat than in the spring.

**Little Bunting** *Emberiza pusilla* A very rare migrant with the only record from outside Sinai coming from Agamy, Alexandria on 2nd November 1980. St. Catherine's Monastery has had 3 records in the 1970s with wintering birds recorded but there were 2 birds at Sharm el Sheikh on 17th October 1991. Eilat has had up to 6 birds in a year with some wintering.

**Yellow-breasted Bunting** *Emberiza aureola* Only three records exist for this species with the first at St. Catherine's on 3rd September 1972 being an adult male and a female/ immature was at Sharm el Sheikh on 17th October 1991 with another at Zaraniq on 3rd September 1993. Up to six have been found at Eilat in the autumn and at least one has overwintered here.

**Reed Bunting** *Emberiza schoeniclus* A rare winter visitor with some very old records. There was a single record from Nile cruises with 2 seen on 17th March 1984 near Beni Suef and a single bird at Al Alamein in 1997.

**Red headed Bunting** *Emberiza bruniceps* A single bird was photographed at Eilat on 22nd May 1979.

**Black headed Bunting** *Emberiza melanocephala* Most records for this species come from Sinai where up to 5 have been seen at St. Catherine's in the autumn. There were 2 males at Ain Musa on 5th May 1990 and a male at Wadi El Arish in April 1992. Away from Sinai there was a single male at El Gadra, Wadi el Natrun on 17th May 1990. It is a common migrant at Eilat with up to 100 each autumn but sometimes peaks at 700 in spring.

**Corn Bunting** *Miliaria calandra* A common winter visitor to the Mediterranean coast which is especially good for this species. It was seen on early Nile cruises as far south as El Til but no records exist for recent cruises. A single bird was seen by visiting birders at Abu Sir pyramids on 31st March 1995.

***Lesser Frigatebird** *Fregata ariel* First for the Western Palearctic at Eilat on 1st December 1997

*Out of sequence

## Bibliography and Reference Books

Although there are few bird books aimed at solely the birds found in Egypt, there are many which help to identify the species and can give an understanding of the bird and its characteristics. There are also many scientific reports, tour guides of birds seen and information on other aspects of wildlife in Egypt, all adding to the information in this book.

**These references are not shown in any particular order.**

*Ornithological Studies in Egyptian Wetlands 1989/90 by Peter L Meininger and A.M.Atta, A WIWO report no 40*

*Finding Birds in Egypt by David Gosney, 1993, Gostours publication ISBN 1 898110 03 4*

*Egypt and the Sudan by Scott Wyne and Damien Simonis, 1996, Lonely Planet ISBN 0 86442 191 5*

*Birds of Egypt by Goodman and Meininger, 1989, Oxford University Press ISBN 0 19 857644 7*

*Egyptian Hotel Guide by the Egyptian Hotel Association, 1996-1997*

*Student's Flora of Egypt by Vivi Tackholm, 1974, Cairo University*

*The Birds of Israel by Hadoram Shirihai, Academic Press ISBN 0 12 640255 8*

*Birds of the Hashemite Kingdom of Jordan by Andrews, 1996, Andrews*

*Birds of Libya by G Bundy, 1976, British Ornithologists' Union*

*Distribution Atlas of Sudan's Birds by G. Nikolaus, 1987, Bonn Zoological Society no 25*

*Environmental Profile Fayoum Governorate, Egypt by Euroconsult/ Darwish Consulting Engineers, 1992*

*The Flora and Fauna of Crocodile Island by Movenpick Hotels [in English and in German]*

*Bird reports from Swan Hellenic Cruises, 77 New Oxford St., London*

*The Birds of Ancient Egypt by Patrick F. Houlihan, The American University in Cairo Press ISBN 977 424 185 1*

*The Catching of Birds in North Sinai, Autumn 1990 by Sherif M. Baha El Din and Waheed Salama, ICBP report no 45*

*The catching of corncrakes and other birds in Northern Egypt by Sherif M Baha El Din, Birdlife International no 55*

*Trapping and Shooting of Corncrakes on the Mediterranean Coast of Egypt by Sherif M. Baha El Din, Bird Conservation International, 1996 6 213 - 227*

*Street Trees of Egypt by M Nabil el Hadidi and Loutfy Boulos, 1989, American University in Cairo Press ISBN 977 424 173 8*

*Common Birds of Egypt by Sherif M. Baha El Din, 1994, American University in Cairo Press ISBN 977 424 239 4*

*Lapwings and Libyous in Ancient Egypt by Alessandra Nibbi 1986 ISBN 010 70428*

*Egypt - Country Study on Biological Diversity by Egyptian Environmental Affairs Agency, 1995*

*The Butterflies of Egypt by Torben Larsen, 1990, The American University in Cairo Press ISBN 977 424 218 1*

*Natural Selections by Richard Hoath, 1992, American University in Cairo Press ISBN 977 424 281 5*

*The Animal World of the Pharaohs by Patrick F. Houlihan, 1996, American University in Cairo Press ISBN 977 424 3617*

*Libyan Mammals by E Hufnagal, 1972, Oleander Press ISBN 0 902675 08 7*

*Fish and Fishing in Ancient Egypt by Douglas J. Brewer and Renee F. Friedman, American University in Cairo Press ISBN 977 424 224 6*

*Raptor Migration at Eilat by Hadoram Shirihai and David A. Christie, British Birds Vol 85 no 4 April 1992*

*Ornithological Winter Observations from Southern Sinai by D. Robel, 1996, Zoology in the Middle East, 12*

*Global Environmental Facility: Egyptian Engineered Wetlands prepared for the United Nations Development Programme by P. Lane & Associates Limited*

*Nicoll's Birds of Egypt by R. Meinertzhagen, 1930, Hugh Rees*
*(Until 1989 the only full work of the birds of Egypt)*

*Geology of Egypt by R. Said [1962] Elsvier*

*Geology and Geography of Egypt by Blebaz Farouk, 1984, QE328E4E57*

## Reports sent to me for this book:-

'Visit to Egypt' April/May 1991 by Marten Ajne and others
'Upper Egypt' December 1992 by Bryan and Laureen Roberts
'Upper Egypt' November 1994 by David Murdoch
'Visit to Egypt' March 1995 by Darrell Clegg and others
'Upper Egypt' May 1995 by Graham Waugh
'Hurghada' 1995 by Roy and Jane Atkins
'Nile Cruise' September 1996 by Chris Tomlinson
'Upper Egypt' September 1996 by Dave Cudmore
'Visit to Egypt' April/May 1997 by Stephen Mawby
'Red Sea diving/birding' December 1995 by Craig and Linda Burnet
'Nile Cruise + Abu Simbel' 1997 by P. Adderley

## Books soon to be published include :-

The new edition of 'Birds of Egypt' (1999+)
'Important Bird Sites in Egypt' (1998?)
'Mammals of Egypt' (1998?)

## The following books help in the identification of the birds:-

'Collins Guide to the Birds of Prey of Britain and Europe, North Africa and the Middle East' by Gensbol and Bertel, 1993, Harper Collins

'Collins Pocket Guide: Birds of Britain & Europe with North Africa & The Middle East' by Heinzel & all, 1995, Harper Collins - covers all the birds

'Birds of Europe with North Africa and the Middle East' by Lars Jonsson, 1992 [paperback 1996] Helm - covers all the birds

'Birds of the Middle East' by Richard Porter and all, 1996, Poyser - Egypt is not mentioned but most of the birds are covered

'Birds of the Middle East and North Africa' by Hollam and all 1988, Poyser - this does cover the birds but there are no plates of the European birds

'Birds of the Western Palearctic'. Handbook of the Birds of Europe, the Middle East and North Africa, 9 Volumes, editors Cramp and Perrins, Oxford University Press - too big to carry with you!

## Publications/organisations which often have articles about Egyptian birds:-

***Sandgrouse*** is the publication of the ornithological society of the Middle East c/o The Lodge Sandy Bedfordshire SG19 2DL, UK. This has the most references of any current publication on Egyptian bird-life and monthly bird records. There was also a small 'bulletin' published alongside the magazine which has now been absorbed by the present magazine.

***Birding World,*** Sea Lawn, Coast Road, Cley-next-to-the-Sea, Holt, Norfolk NR25 7RZ, UK.

***Birdwatch,*** Solo Publication Ltd, Bow House, 153 - 159 Bow Rd London E3 2SE, UK.

***Birdwatching,*** EMAP, Bretton Court, Bretton, Peterborough, PE3 8DZ, UK.

***Dutch Birding,*** PO Box 737, 9700 As Groningen, The Netherlands.

***Courser*** was the publication of the Egyptian Bird Club which sadly no longer exists.

**World working group on birds of prey and owls,** Wangenheimstr.32, 14193 Berlin, Germany. Fax 49 30892867. Currently working on radio tracking many species of raptors passing through Egypt.

## Maps:-

'Road Atlas of Egypt' by Shell Marketing and the Automobile Association UK,1996 - still many birding roads missing! eg No water in Wadi el Rayan and no road to it!
'Egypt' by Kummerly and Fry, 1995/96. Some maps are reprints of older maps sold

in Europe whereas if bought in Egypt they are more up-to-date eg Wadi el Rayan still has no water and many western areas are missing with Gebel Elba also!

## A useful addition:-

BBC Arabic Phrase Book by Nagi el Bay & Victoria Floyer-Acland, 1996, BBC Books ISBN 0 563 39996 1

## Bird Information Services:-

Dutch Birding Travel Reports Service, PO Box 737, 9700 As Groningen, The Netherlands.
Bird Reports from around the World, Steve Whitehouse, 6 Skipton Cresent, Berkeley, Pendesham, Worcestershire, WR4 0LG, UK.

## Egyptian bird sightings, radio-tracked birds and bird reports:-

http://www.netlink.co.uk/users/ag/osme/osmehome.html
Type in 'stork migration map' on the Internet [griffon vulture and steppe eagle included]

**Sinai Wildlife Centre**, Coral Bay, Sharm el Sheikh Tel Sharme 062 601610

## Companies running birdwatching trips to Egypt at the present time (1997):-

*Bird Holidays*, Mantra WGT Ltd, Oxford House, Oxford Road, Guisley, Leeds, Yorkshire LS20 9AA, UK

*Birdfinders,* 18 Midleaze, Sherborne, Dorset, UK DT9 6DY,

*Sunbird,* Box 76, Sandy, Bedfordshire, SG19 1DF, UK

*The Travelling Naturalist*, 9 Little Britain, Dorchester, Dorset, DT1 1NN, UK

*Branta,* 7 Wingfield, St, London SE15 4LN, UK

*Birdquest,* Two Jays, Kemple End, Birdy Brow, Stonyhurst, Lancashire, BB7 9QY, UK

*Limosa Holidays,* Suffield House, Northrepps, Norfolk, NR27 0LZ, UK

*Swan Hellenic Cruises,* 77 New Oxford St, London, WC1A 1PP, UK

## Companies based in Egypt running tours:-

This can be anything from full Cairo to Aswan Cruises [Thomas Cook] to guided walks, day/half day tours added on to present schedules.

*Thomas Cook,* 12 Maiden el Sheikh Youssef, Garden City, PO Box 165-11511, Cairo, Egypt. Tel 3564650/2 -3564934/6, Fax 3545886

*Abercrombie & Kent Egypt,* 18 Youssef el Gendy St. , Boustan Centre, Cairo, Egypt. Tel 3936255 Fax 3915179

**The Author:** John Miles was born in Yorkshire, England and brought up in this county. He left school to work outdoors in forestry and then in wildlife protection which took him, not only around the UK, but also to many countries of the world. He worked for the RSPB (The Royal Society for the Protection of Birds) from 1976 to 1991, after which time he returned to his present home in Cumbria with his wife and three sons.

It was in Cumbria that he decided to add writing to his work as a wildlife consultant, in the form of two books, 'Hadrian's Birds' and 'Exploring Lakeland's Wildlife'. In addition, he has contributed many articles to wildlife magazines and to the media in general. He continues in his plight to help the birds of prey in Britain, which are facing an uncertain future due to a lack of leadership by the governments and bird societies, and is a strong campaigner for the greater understanding of the needs of the British Uplands. John has added many scenic and bird pictures to this book.

**The Artist:** Mike Henry was also born in Yorkshire and is presently Principal Teacher of History at Dame Allan's School in Newcastle-Upon-Tyne, England. He was also the artist for John's two previous books, and is likewise a campaigner on behalf of the birds of prey in Britain.

**The Photographer:** Keith Temple also started life in Yorkshire moving up to Cumbria like John and enjoys the challenge of obtaining great bird photos. He spends many hours photographing birds around Cumbria and the world - many of his shots have appeared in bird magazines.

## Acknowledgements

### I would like to thank sincerely the following people:-

Adel Rady and Douglas Graham for making this book possible.

All the members of staff at the World Bank and Tourism Development Authority for their help and co-operation.

Dr Salima Ikram and Dr Viv Callender for allowing me into their world of Egyptology.

Richard Hoath for his many tales of birds around Egypt and his efforts to make a future for them.

Sherif and Mindy Baha El Din for their constant efforts to help Egyptian wildlife.

Both the hotel staff and the guests at the Garden City House Hotel, Cairo for keeping me informed about the world.

The many people in Egypt who made my stay so enjoyable, not forgetting Mahemoud Sayd, our driver around Egypt, and Hehab from Cairo.

The Swan Hellenic staff and crew who made my journeys flow so well along the Nile.

Mike Henry for supporting my books with his talents yet again.

Keith Temple for his great work with the lens.

Andrew Hetherington for his encouragement and slides, plus all those helping with information/articles about Egypt whose names do not appear in print.

Not forgetting my wife, Thelma, and my boys Jamie, Tom and Ewan, for letting me be away for so long, and for their support once I had returned.

Abu Simbel Temple